Local Voices, Local Choices

Local Voices, Local Choices

The Tacare Approach to Community-Led Conservation

Created by the Jane Goodall Institute

Introduction by Jane Goodall

Foreword by Jack Dangermond

Edited by Lilian Pintea and Adam Bean

Esri Press
REDLANDS | CALIFORNIA

Published and distributed under exclusive license to Esri Press.

Esri Press, 380 New York Street, Redlands, California 92373-8100

26 25 24 23 22   1 2 3 4 5 6 7 8 9 10

ISBN: 9781589486461

Library of Congress Cataloging-in-Publication Data

Names: Jane Goodall Institute, author.
Title: Local voices, local choices : the TACARE approach to community-led conservation / created by Jane Goodall Institute.
Description: Redlands, California : Esri Press, [2022] | Includes bibliographical references.
Identifiers: LCCN 2021056778 (print) | LCCN 2021056779 (ebook) | ISBN 9781589486461 (hardback) | ISBN 9781589486478 (epub)
Subjects: LCSH: Community-based conservation. | Conservation of natural resources--Citizen participation. | Conservation projects (Natural resources)
Classification: LCC S944.5.C57 J36 2022 (print) | LCC S944.5.C57 (ebook) | DDC 333.72--dc23/eng/20211202
LC record available at https://lccn.loc.gov/2021056778
LC ebook record available at https://lccn.loc.gov/2021056779

Printed in the United States of America. This book is made from Forest Stewardship Council® certified paper.

Contents

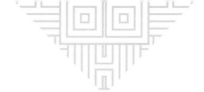

Foreword

It's a rare person who can make significant scientific discoveries, help redefine the relationship between humankind and the natural world, inspire millions of young people to make a positive difference in the world, and at the same time work to break down gender barriers and successfully advocate for widespread change—all while becoming a household name. I'm referring of course to Dr. Jane Goodall.

As a young child, she dreamed of living in and exploring the geography of Africa. In 1960, Jane took the initiative to live that dream and, as a young woman, set off to Gombe, in what is now called Tanzania. She has studied chimpanzees there for more than 60 years, and what she found not only defined the first half of her career but also redefined humankind, shattering the myth that humanity is somehow distinct from and superior to nature and our closest ancestors within it.

Jane is an extraordinary person who has accomplished so much. Her science and advocacy for conservation of nature have impacted the hearts and minds of millions of people. In recent years, she has taken what she learned from her early experiences working with chimpanzees and the communities around Gombe and worked on messages and strategies for creating sustainable communities that are adjacent to and integrated with natural areas. This book shares and brings together some of her key thinking and approaches.

Changing geography, changing world

A combination of recent trends—human population growth, careless land development, and climate change—is threatening the survival of many living species, including the human species.

Jane saw this threat herself when she returned to her beloved Gombe in the 1990s and noticed the overwhelming impact that humans were having on this ecosystem. This impact was not just threatening the existence of the chimpanzees but also the health and well-being of the people who were degrading the landscape as they struggled to survive. She realized that to improve the lives of animals, she must first figure out how to improve the lives of humans. It was not enough to create a plan for protecting biodiversity; true sustainability required creating a landscape-level planning process that engaged the local community and reflected its values and needs.

Today, humanity finds itself in an odd position. We realize that, without intending to, humans are collectively causing massive environmental degradation and destruction of the natural world. At the same time, we hold in our hands the keys to halt and reverse these trends. So how can we move away from unsustainable actions and toward developing sustainable livelihoods for both communities and the natural world on which we all depend? That's what this book is about.

Taking care

The second half of Jane's career has been defined by promoting advocacy, improving educational outreach, and developing messages and strategies for sustainable community development. She has successfully taken the skills and techniques learned in the first half of her career and broadened her impact through the creation of new outreach programs such as Roots & Shoots and the Lake Tanganyika Catchment Reforestation and Education (TACARE) program.

The Jane Goodall Institute (JGI) launched TACARE in 1994. Now known simply as Tacare, this community-centered conservation and development program partners with local people to create sustainable livelihoods while promoting environmental protection. The Tacare approach integrates geographic information (maps, aerial images, and data) with collaborative design thinking involving all the stakeholders in a local community. Tacare achieves conservation results by consulting closely with communities about their priorities, recognizing the critical need for local community input and engagement to help restore and reconnect fragmented natural areas critical to biodiversity.

Jane's work and the work of her colleagues featured in this publication illustrate a viable pathway to a more sustainable future, where both humans and the natural world can thrive together.

Changing the world together

Tacare is Jane's vision of how we can collaborate to make the world a better place. She has learned from decades of experience in the field that this vision can only become reality when the cultural, physical, and biological aspects of geography are brought together to holistically inform community thinking and action.

Jane and I share a deep belief in the interconnectedness of all things and the obligation of humankind to protect and conserve our environment. Tacare is an approach that includes planning, policy, and technology working hand in hand to take care of the geography around us. It is grounded in empathy and based on working together as peers, regardless of the privileges bestowed on many of us in the field of conservation.

This book encapsulates the work of Jane and her colleagues and articulates a fundamental approach that will be necessary for the future. The stories in these pages are told by local conservation practitioners who are achieving sustainable landscape-level conservation.

Studying and applying this approach can help all of us make positive impacts in communities everywhere, as well as in the natural world.

—Jack Dangermond
Founder and president of Esri

Acknowledgments

Tacare reflects the commitment and close collaboration of numerous people and institutions over the years. Their support and critical feedback have helped shape the concept and fostered its growth and evolution into the community-led approach it is today. Their influence cannot be overstated.

Beginning with those who developed the original concept and proposal and helped secure the initial funding, we thank Jane Goodall, Anthony Collins, Dilys McKinnon, Tim Clarke, Garth Bowman, and Janette Wallis.

Our heartfelt gratitude goes to all members of the Tacare team in Kigoma, Tanzania, along with those of the Gombe Stream Research Center, Roots & Shoots, and the broader JGI-Tanzania staff who, at various intervals since 1994, have developed Tacare and acted as envoys within the local communities. These include George Strunden, Emmanuel Mtiti, Aristides Kashula, Mary Mavanza, Freddy Kimaro, Shadrack Kamenya, Amani Kingu, John Mike, Sood Ndimuligo, Karen Zwick, John Maclachlan, Grace Gobbo, Deus Mjungu, and Paul Cowles. We also thank Sania Lumelezi, Fadhili Abdallah, Jovin Lwehabura, Paul Mjema, Elikana M. Manumbu, Phoebe Samwel, Jumanne Kikwale, Rashidi Kikwale, Hilali Matama, Gabo Paulo Zilikana, Eslom Mpongo, Hamisi Mkono, Yahaya Almas, Ashahadu Jumanne, Baraka Mussa, Japhet Mwanang'ombe, and Merlin van Lawick.

Over the years, Tacare was adapted and refined by integrating new approaches, technologies, and tools, thanks to the feedback, insights, and support from numerous experts, including Lilian Pintea, David Gadsden, Michael Wilson, Anne Pusey, Lynne Gaffikin, Elizabeth Lonsdorf, Dominic Travis, Terry Cook, Robert Sassor, Elizabeth Gray, Michelle Brown, Margo Burnham, Caroline Byrd, Gwynn Crichton, Ryan Luster, Matt Brown, Tim Tear, Cristina Lasch, Nick Salafsky, Gerald Kinn, Shannon McElvaney, Matt Artz, Mike Ruth, Tanya Birch, David Thau, Rebecca Moore, Chuck Chaapel, and Richard Wrangham, among others.

We would like to thank the JGI staff and board members who supported Tacare over the years, including David Shear, James Lembeli, Keith Brown, Alice Macharia, Carol Collins, Eduardo Hernandez, Adam Bean, Mary Paris, Bill Wallauer, Dan DuPont, Shawn Sweeney, Ashley Sullivan, Carol Irwin, Mary Lewis, Susana Name, Robert Eden, Don Kendall, Steve Woodruff, and Reed Oppenheimer.

Our ongoing gratitude goes to local communities, governments, and JGI chapters in Tanzania, Uganda, the Democratic Republic of the Congo (DRC), the Republic of the Congo, Senegal, and other countries that adopted Tacare in their respective countries and programs to meet local needs. A special thanks to Peter Apell, Dario Merlo, Rebeca Atencia, Debby Cox, Timothy Akugizibwe, Robert Atugonza, Fernando Turmo, Liliana Pacheco, Federico Bogdanowicz, Diana Leizinger, and many others, whose contextual adaptation of Tacare has broadened the outreach of Tacare-type community-led conservation.

The US Agency for International Development (USAID) played a key role in funding, supporting, and scaling up Tacare since 2005. Tacare work with the local communities was supported by generous funding and in-kind contributions from partners, including the Packard Foundation, Wanda Bobowski Fund, European Union, Oppenheimer Brothers Foundation, United Nations Development Programme (UNDP), United Nations International Children's Emergency Fund (UNICEF), Esri®, Maxar, Planet, Google Earth Outreach, Microsoft, Esri Eastern Africa, Blue Raster, National Aeronautics and Space Administration (NASA),

Global Forest Watch/World Resources Institute (WRI), Arcus Foundation, and US Fish and Wildlife Service.

For their continued support, we would like to thank Jack and Laura Dangermond, Nancy and James Demetriades, George Macricostas, and Loretta and Chris Stadler.

And finally, we would like to acknowledge the power of shared vision, specifically that which exists between JGI founder Jane Goodall and Esri founder Jack Dangermond, for it was this that made this book possible.

From the editors

I remember when I was first introduced to the TACARE project by Jane Goodall, Emmanuel Mtiti, and George Strunden back in 2000 in Kigoma. It has been a privilege to work closely with the Jane Goodall Institute team, local communities, governments, universities, and many other partners for the last 20 years and play a role in the evolution of Tacare, from a project to a community-led approach that could be scaled across the world. I am very grateful to the generosity of Jack and Laura Dangermond, who made this book possible. Finally, I would like to thank my spouse, Anna, my son, Codrut, and my daughter, River, for their support during my numerous trips to sub-Saharan Africa and for keeping us connected and inspired as part of our own communities back home in Silver Spring, Maryland.

—Lilian Pintea

It is with immense gratitude that I acknowledge Dr. Shadrack Kamenya and Mzee Jumanne Kikwale of JGI-Tanzania and Dr. Peter Apell and Timothy Akugizibwe of JGI-Uganda for providing critical cultural guidance during the research phase of this publication. Their personal relationships with local community members, built on years of trust and mutual respect, made the interviews, translation, and contextual understanding possible, allowing local voices to be captured herein.

—Adam Bean

Special thanks

Esri Press thanks Matt Artz and Jenefer Shute for their special editorial help.

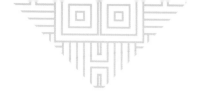

Introduction

The origins of Tacare

The Jane Goodall Institute's approach to community-led conservation

Many people are still surprised when they realize that I, "the chimpanzee lady," have for years been working on a variety of conservation initiatives that have forced me to leave the chimpanzees and the forest to work with local communities and to travel around the world—raising awareness about the threats to both chimpanzees and people in Africa, and the effect of hundreds of years of human exploitation of the planet's natural resources. In a way, it all started when I was an animal-loving child, who spent hours out in nature watching birds and squirrels and insects in my hometown of Bournemouth, in England. I had a wonderful and supportive mother who encouraged my interest, finding books about animals from the library. When I was 10 years old, I decided I would go to Africa, live with wild animals, and write books about them. This was in 1944, when girls simply did not do things like that; anyway, we had very little money and Africa was relatively unknown to outsiders. Everyone told me I should dream about something I could do—except my mother, who told me that I would have to work very hard, take advantage of every opportunity, and then, if I did not give up, perhaps I would find a way.

As is well known, I did get to Africa, arriving in Kenya in 1957. It was the famous paleoanthropologist Dr. Louis Leakey, who made my dream

Jane Goodall. *The Jane Goodall Institute, Bill Wallauer.*

come true: he found money for me to go and study the chimpanzees of what is now Gombe National Park on the shores of Lake Tanganyika (amazing as I had not even been to college, since we could not afford it). Back then it was the British protectorate of Tanganyika, once part of German East Africa but taken over by the British after World War I. It became the independent Republic of Tanzania in 1961, just one and a half years after I had arrived in Gombe.

Louis Leakey found it hard to get funding for what was then considered a crazy idea—sending a young woman into the forest—but he eventually got funding for six months from an American philanthropist. But then the authorities refused permission for me to go to the forest; they did not want to accept the responsibility. Louis, however, persisted and they eventually agreed—provided I took a companion. It was my amazing mother who volunteered to come. We shared one secondhand ex-army tent and lived mostly out of tins along with rice and local beans prepared by our cook. Mum played a very important role. In the early days, the chimpanzees vanished into the forest whenever they saw me, and I got worried that I wouldn't find out anything significant before the money ran out. End of study, end of dream. But Mum pointed out all the things I was learning as I spent every day, from dawn to dusk, searching for chimps and watching them in the distance through binoculars. Even more important, she set up a little clinic, with very simple medications like aspirin, Epsom salts, and saline drips. She made some remarkable cures, and so, right from the very beginning, we had excellent relations with the fishermen from the villages

Jane Goodall (*left*) with the archaeologist and paleoanthropologist Dr. Louis Leakey in Kenya, circa 1957. It was Jane's attention to detail, observational skills, and, owing to a lack of formal scientific training, her unbiased and curious mind that prompted Louis to support Jane's dream of studying wild animals. *The Jane Goodall Institute, Joan Travis.*

along the lake shore. They began coming from farther and farther away to see Mum—I found out later she was known as the White Witch Doctor.

It was just after Mum had left to return to England that my luck turned. By then one chimpanzee—a handsome male I had named David Greybeard—had begun to lose his fear of me and, on this never-to-be-forgotten day, I saw him using grass stems to "fish" for termites. And sometimes he picked leafy twigs and had to remove the leaves before he could use those as tools. He was not only using but making tools, something that scientists believed was a behavior unique to humans. In fact, we humans had come to be known as "Man the Toolmaker." It was this discovery that brought the National Geographic Society into the story. They not only provided funding so that I could continue my study, but sent Hugo van Lawick, a photographer and filmmaker, to document the first

Zinda, a Gombe chimp, uses a modified stem to fish for termites. This behavior was Jane's first breakthrough in observational research and the first time Western science was confronted with hard evidence of toolmaking and tool use among animals. *Nick Riley, 2010.*

study of the behavior of wild chimpanzees, who are, along with bonobos, our closest living relatives. Gradually I got to know the various chimp individuals and their complex society, and the many ways their behavior resembles our own.

Soon Hugo and I began a small research station with student volunteers to help collect information. We began employing local Tanzanians from the surrounding villages to help with the research. They got to know the chimpanzees and followed them through the forest, first with the volunteers but then on their own. I believed that once they realized how amazing these chimpanzees are, they would share the information in their villages and this would help people better understand these humanlike neighbors of theirs and thus prevent poaching. And, indeed, this proved to be the case.

How a chimpanzee study led to Tacare

In the late 1980s, I flew over the tiny 14-square-mile (about 36-square-kilometer) Gombe National Park and the surrounding area in a small plane. I was deeply shocked. When I began the research in 1960, the area was part of the forest belt that stretched across equatorial Africa to the west coast. But now I looked down on a small oasis of green forest—the park—surrounded by bare, treeless hills.

It was clear that, thanks to high population growth, there were more people than the land could support. And their numbers were swollen by refugees from the conflicts in Burundi and the Democratic Republic of the Congo (DRC). The people were too poor to buy food elsewhere; their land was overfarmed and largely infertile. Women had to walk farther and farther from their villages in search of wood for fuel, adding hours of labor to their already difficult days cooking for their large families. Looking for new land to clear for their crops, people had turned to ever steeper and more unsuitable hillsides. With the trees gone, soil was washed away during the rainy season, causing bad erosion and frequent landslides. The streams that originate from the Rift Escarpment watershed and empty into Lake Tanganyika had become increasingly silted.

All of this meant that the chimpanzees were more or less trapped within the tiny national park, cut off from other groups. There could be no exchange of females between groups—which prevents inbreeding—and with only some 100 individuals remaining, the long-term viability of the Gombe population was at risk. Yet how could we even hope to protect them while the people living around the borders were struggling to survive, envious of the lush, forested area from which they were excluded? That's when it hit me that unless we could help the people find ways of making a living without destroying their environment, we could not hope to protect chimpanzees, their forests, or anything else. And so the idea for Tacare began.

The first discussions were with Garth Bowman [a UK farmer who worked with Kigoma fishermen to reduce the rate of erosion], but when he had to return to Europe to school his children, George Strunden took over. He recruited a small team of local Tanzanians who had worked with nongovernmental organizations (NGOs) in agriculture, forestry, water, and health issues to visit the 12 villages closest to Gombe. To listen to the people, learn from them about their problems, and find out how they felt, the Jane Goodall Institute (JGI) could help. We learned that, at that time, their main concerns were the need to grow more food and have better access to primary health care and better education facilities.

In consultation with village leaders, George and his team developed a holistic plan to address these needs and at the same time help improve the lives of the villagers in an environmentally sustainable way. Conservation of wildlife was not mentioned; the villagers were already resentful that Gombe had been set aside for chimpanzees, and it was first necessary to assure them that we truly cared about their welfare. But we believed (correctly!) that our approach would ultimately help protect the chimpanzees, as well.

First we had to obtain funding to put the plan into action. Once a proposal had been drawn up, I wanted, with Dilys McKinnon, the ED of JGI-UK, to try our luck with the European Community (EC), now the European Union (EU). Tim Clarke, whom we knew well from his years working for the EC in Tanzania, was working as the focal point for environment in the Directorate General for Development, responsible for managing the EC's €100-million-per-year environment budget. This was amazingly lucky for us: JGI was an unknown organization, and Tim was undoubtedly helpful in securing that first grant as he knew JGI in Tanzania. But even though Tim subsequently told me that our approach to community engagement was innovative and scored lots of points, we were nevertheless forced to "focus." We were told we could not do everything. So we applied to their forest conservation fund, which refused to fund non-forestry-related activities but gave us funding for three years to pilot the Lake Tanganyika Catchment Reforestation and Education (TACARE) project, mainly intended to establish tree nurseries in the villages along the Rift. We had to fundraise with other donors to implement our holistic, community-focused approach: from the United Nations Development Programme (UNDP) for agriculture, the Rabobank Foundation for microfinance, and United Nations Children's Fund (UNICEF) for water and environmental sanitation. And later we got money from the Packard Foundation for family planning and the Wanda Bobowski Fund for girls' scholarships.

George, who had been working in agriculture for 15 years in that part of Tanzania, knew the importance of getting as many villagers involved as possible. He put together a small group of women who accompanied the team to the villages and enacted short skits about the significance of the environment. They were accompanied by a few local musicians and dancers. This was very popular, as visits from outside were rare in those days. The JGI team was welcomed, and good relationships with these communities were cemented.

Eventually, we were able to introduce terracing, showing the villagers how they could restore fertility to overused soil (without the use of fertilizers), and we started improving health and education facilities in the villages by collaborating with the local Tanzanian authorities. And, as the villagers came to trust JGI, and once we had obtained funding, we were able to introduce some of the other projects such as water management, microcredit, and agroforestry.

Gradually our work extended to other villages. The original project description was no longer appropriate, and our project is now known simply as Tacare (pronounced "ta car eh") and stands for "take care"—it takes care of people, the environment, and animals. As additional funding became available, we were able to extend inland through the greater Gombe ecosystem (GGE) and the huge area south of Gombe, the Masito-Ugalla part of the greater Mahale ecosystem (see map on overleaf).

The Tacare model proves successful

Tacare, as the following chapters show, has become one of the most comprehensive community-led conservation programs in Africa, designed to

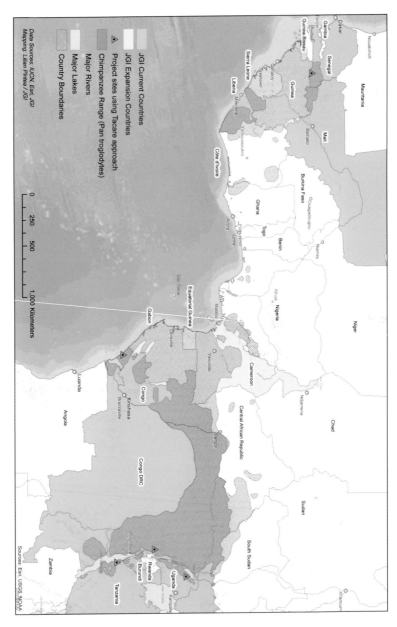

The African countries in which JGI operates, with locations of Tacare projects in relation to chimpanzee range habitats. *The Jane Goodall Institute, Lilian Pintea.*

address poverty and support environmentally sustainable livelihoods. It has developed into a powerful, holistic program that restores fertility to overused farmland (without the use of agricultural chemicals) and offers training in improved farming and agroforestry practices, water management, and marketing skills. We work with government agencies to improve primary health care and children's education. We provide microcredit opportunities, particularly for women, based on the principles of Muhammad Yunus's Grameen Bank, and scholarships to give girls a chance for secondary education. And this, I discovered, meant building improved pit latrines with separate facilities for boys and girls and providing girls with sanitary towels. Volunteers from the villages attend workshops, learn about family planning, and offer this information to village families. These services are well received because there is growing understanding that a good education is a way out of poverty, and families cannot afford this for the 8 to 10 children that used to be the norm.

Under the leadership of Dr. Lilian Pintea, we have employed cutting-edge geospatial mapping technologies (with support from Esri®, Maxar, Planet, Google Earth Outreach, and NASA) to produce high-resolution maps, helping the villages to create the land use management plans required by the government and to monitor success. In their land use management plans, the villages around Gombe have set aside areas to form buffer zones between the park and the villages, reducing the potential for conflict between humans and wildlife.

Conservation

Thus, it was with the support of local communities and government officials that we were able to integrate planning for the conservation of chimpanzees and their habitat into the Tacare model. Using scientific data and local knowledge, we gained greater understanding of what was needed to protect the chimpanzees and find ways to address the key threats to them and their habitats. In this, the cooperation of the villagers has been crucial. Most of Tanzania's remaining chimpanzee populations do not live in the protected areas of Gombe and Mahale Mountains National Parks but in village forest reserves and recently established local authority forest reserves that are managed by district governments and local communities.

Over the past 30 years, many of the trees in these areas, growing from seeds and tree roots left in the ground, have reached heights of over 20 feet. Other villages have set aside land for reforestation outside Gombe National Park that will form contiguous stretches of forest habitat, acting as corridors that enable the previously isolated Gombe chimpanzees to interact with other remnant groups and prevent inbreeding. Several females have already made use of these corridors, which are also used by other wildlife, thus greatly benefiting the biodiversity of the area.

With support from the United States Agency for International Development (USAID) and other donors, we now work in 104 villages in the greater Gombe ecosystem, and the large contiguous area in the south, the Masito-Ugalla ecosystem. In this area, we have funded many of the government-required village land use plans and trained volunteer forest monitors from the villages to patrol their village forest reserves. They use mobile apps on smartphones to report on the health of their forests, pinpointing illegal activities (such as tree cutting or animal traps), the success of forest restoration efforts, and sightings or indications of the presence of animals such as chimpanzees, leopards, and pangolins. All this information is collected in a standardized way and sent immediately to Western Tanzania Decision Support and Alert System, managed by ArcGIS® Online in the cloud. JGI staff, local officials, and other partners can use dashboards to download and view it for analysis or share it with decision-makers in the village and district governments. Thus, the information is totally transparent. The total area covered by our various projects in western Tanzania is now almost 6,732 square miles (17,435 square kilometers).

Education: Roots & Shoots

A final and important component of Tacare is our environmental and humanitarian program for youth, Jane Goodall's Roots & Shoots. From the very beginning, we introduced this program into the village schools. It is a program for all ages and encourages children to choose projects to improve the lives of people, animals, and the environment. The children plant trees, grow organic food in school gardens, and learn about wildlife and the importance of preserving biodiversity. All the effort we put into protecting habitats and improving lifestyles will be useless unless new generations grow up to be better stewards of the planet than we have been.

The success of the Tacare approach

The Tacare approach has a proven track record. We have achieved successful partnerships with hundreds of thousands of villagers, improved their health and education facilities, provided family planning information, empowered women to take a stronger role in village life, and encouraged villagers to respect and protect their environment and engage in environmentally sustainable livelihoods. We have acquired a sophisticated understanding of socioecological and political systems that has enabled us to work harmoniously with farmers, village leadership, and local and central government authorities. We have learned how to combine innovative mapping technologies and local traditional knowledge and integrate science into local decision-making processes to develop smart conservation and land use plans. In this way, we have been able to encourage villages to adopt new, improved agricultural and market practices while simultaneously protecting the environment—the broader landscape they rely on for long-term survival—and to conserve and restore the all-important forest ecosystems with their rich biodiversity.

JGI has also developed Tacare-type programs in Burundi, Uganda, the DRC, the Republic of the Congo, Senegal, and Guinea, and a similar program is being developed in Mali.

It is with great pride I introduce you, the reader, to this book. It captures 30 years of people, stories, trials, and successes as the Tacare approach has evolved since 1994. Most of the individuals telling the stories in this book are close friends, students, and mentors in the complex world of community-led conservation. Together we have created a program of which all participants—past as well as present—can be justly proud. It is a program that embraces the conservation of chimpanzees and other wildlife and the health and education of whole communities, and it is, I believe, the very embodiment of hope for the future of our planet.

—Jane Goodall, PhD, DBE
Founder, the Jane Goodall Institute, and United Nations Messenger of Peace
www.janegoodall.org

Chapter 1

The human-made island

Jumanne Kikwale meets Jane Goodall at an impressionable age

Anthony Collins arrives to study Gombe's baboons

When we think of earth's wild frontiers, we usually visualize vast stretches of untamed wilderness with animals running free—herds of wildebeest crossing the Serengeti, colorful fish dancing through the coral of the Great Barrier Reef, massive mountain gorillas moving through the dense understory in the Virunga Mountains. Yet these sweeping cinematic images, spectacular as they are, tell only part of the story. What they often don't show is that these animals are marooned on a speck of finite space, within the boundaries of a national park, protected reserve, or similar designated area.

Often completely encircled by human habitation, the world's protected parks and reserves give the illusion of an infinite wildlife domain, yet they contain approximately just 20 percent of the world's biodiversity. The remaining 80 percent is found outside these protected areas, in areas colonized and often overused by people. The most ecologically significant spaces exist in between reserves—spaces such as the margins of protected parks, where the human–wildlife interface represents a perpetual tug-of-war over land and resource use. Compounding this existential standoff is the mosaic sprawl of villages, towns, and cities that block

Jumanne Kikwale (*left*) and Dr. Anthony Collins. *The Jane Goodall Institute.*

habitat connection. Paradoxically, these rich areas of biodiversity usually harbor the poorest of the world's human population.

In lands where wealth is measured by dry firewood and daily access to water, the people living in remote communities on the very fringes of wilderness are isolated from even the most rudimentary of social infrastructures and devote most of their waking moments to securing the basics needed to survive. Time spent trekking to collect water can consume half of their daylight hours, while gathering food and firewood for cooking occupies the remainder. These are the people who do not have the luxury of choice. They do not have the freedom to ponder nutritional preferences or brand decisions at the grocery store, and they do not have the privilege of basic health-care services to treat a snakebite or set a broken leg that so many other Africans enjoy.

What these isolated communities of Africa do hold, however, is a critical key to rehabilitating much of the surrounding ecosystems, a key to increasing nature's chance at recovering from the systematic degradation of habitat and species decline. The key is their proximity to nature. Indeed, about 65 percent of all global land is under indigenous or local community ownership. This is the space in which the Jane Goodall Institute (JGI) has been working to hone its Tacare approach. The custodians of these critically important and biodiverse landscapes are being empowered to take up the mantle for positive, sustainable change. By integrating science and technology with their local knowledge and cultural practices, Tacare practitioners are engaged in an ongoing effort to strike a balance between humanitarian and environmental needs. This is the space where the seeds of change are beginning to take root.

For rural and impoverished regions, sustainability cannot be an either–or choice. A choice placing the focus completely on species or habitat conservation is just as detrimental to basic humanitarian needs as an exclusive focus on human rights is to the surrounding ecology. But humans *are* a part of nature. The two are inextricably linked, and choosing to help just one is to devalue—and potentially doom—the other. Just as every natural process is shaped by positive and negative feedback cycles, JGI's Tacare approach has grown and adapted through cause and effect, starting when a young English woman connected the needs of local people to the needs of her research subject, the chimpanzee.

Two individuals who have been part of this journey from its earliest days are Mzee Jumanne Kikwale and Dr. Anthony Collins. Their decades-long association with JGI is remarkable in an era of nomadic professionals, and it is through their knowledge and longstanding relationships with the people of western Tanzania that the slow yet steady wheels of positive change keep turning.

Named after the Swahili word for *Tuesday*, Jumanne is the longest-affiliated Jane Goodall Institute employee. From among the youngest to meet Jane in 1960 to a wise and well-respected tree in the Jane Goodall forest today, Jumanne is a testament to how powerful a holistic approach can be for a young mind.

"At the time she arrived," Jumanne recalls, "she met my father who was working with the game scouts. They were the people who were guarding the forest. When Dr. Jane came, she asked him to work with her." Then, with pride, he continues, "He was the first guy to take her in the forest to track chimps: his name was Mzee Rashidi Kikwale." The Swahili word *Mzee* is an honorific for a male elder, and often precedes the name of a respected community member. Indeed, Jane has since fondly reflected on her time with Rashidi, and the moment when, under his guidance, she glimpsed her first wild Gombe chimpanzee.

Today, a respected Mzee himself, Jumanne walks through Gombe's surrounding villages with a patriarchal air, weaving through the people to greet old friends, calling out to street vendors with what seems to be cheeky humor, and correcting the behavior of errant children as they swarm and threaten to disrespect the adults. His demeanor is calm and gentle, yet somehow also firm, as one would expect to see of a retired sports coach or former schoolteacher, traits that he showed early on, when he first met Jane in Gombe.

"When they first arrived, Dr. Jane was with her mother," he says, "and when Dr. Jane went in the forest, her mother would stay back in camp, trying to build good relationships with the people. The people around there were fishermen and most used to come to the camp every day to ask something." Jumanne was intrigued by the sheer abnormality of what Jane herself would later describe as "these strange white creatures" arriving on the shores of Gombe. While Jane took her daily outings with Mzee Rashidi in search of chimps, young Jumanne became an almost permanent fixture in camp, assisting Jane's mother, Vanne, with the rudimentary health clinic.

The clinic proved essential to building trusting relationships that would add to the foundation of what is now the longest-ever continuous nonhuman research program in scientific history. Even at such a young age, Jumanne would take on the role of instructor, distributor, and security—as Jane explains, "He'd help my mother identify guys who queued up for her medicine and would say, 'He's already received some, he's just playing games now,'" ensuring they did not sneak through for a second round of treatment. Motivated only by his willingness to help, and the odd free adhesive bandage on a minor scratch, he was growing under Jane and Vanne's positive influence as a young educator and conservationist.

Young Jumanne (*left*) and his father, Rashidi, gather fish on the shores of Lake Tanganyika. *The Jane Goodall Institute, Hugo van Lawick.*

"The best thing she did was to start that small clinic for them," says Jumanne passionately. "When they were sick, feeling malaria or fever, or if they had an injury, they would come to [Vanne] and she would help by giving them *dawa*," the Swahili word for *medicine*, "and sometimes clean their wounds. This made the people friendly towards her, and it was the best way to develop trust." As a boy, Jumanne unknowingly became part of what would later be defined as the early stages of the Tacare approach, unifying conservation with community development. "Before Jane arrived, I didn't know anything about the chimps, and as time went on, I began to realize something. Chimps are creatures that need to live as well as we do, and chimps can't live in a bare place. If you want to have chimps, you must have a forest, but we also need a house, we need food, and other things. All of these things come from trees."

This is part of the disconnect Jane often refers to. The fact that other species rely on the same resources humans do doesn't often occur to people when their sole focus is to secure these resources for their own survival,

Vanne Goodall (*in blue*) dispenses medicine to villagers in Gombe. *The Jane Goodall Institute, Hugo van Lawick.*

or, in the developed world, for their own greed. Linking resource needs with those of neighboring species is a critical component of connecting people to their ecosystems. "I realized this," Jumanne continues, "when she was going in there every day, that chimps needed to be helped by people. I learned that every creature needs a place to live. Just as we [humans] need a house to stay in, animals need the forest to survive. We had to do something for them."

The realization of this connectedness had a lasting impact on Jumanne and set him on the lifelong path of educating others. After his initial experience with Jane, he left to go to school. Then, in 1974, he came back for a year and worked with Jane in Gombe on the baboon research project. This would not be the last time Jumanne would circle back to JGI following the pursuit of education; next time, he would go on to spearhead one of JGI's most influential programs and, for the first time in JGI history, one not based in the wildlife research arena.

It was shortly before Jumanne's return to Gombe in 1974 that another long-term associate emerged. With the ink still drying on his zoology degree, Dr. Anthony Collins arrived in Gombe from Scotland in 1972, with a strict focus on research and animal behavior. Today, on the eve of his fifth decade with JGI and Gombe National Park, Anthony, or Anton as he is known, represents an ocean of knowledge about Jane, the institute, and the history of western Tanzania, its people, and wildlife. "I came for the research program," he says, "but not on chimpanzees. I was researching baboons at the time, which is like their companion species in the forest."

At any given time of year, Anton can be found in Gombe National Park, where his office is based. As he walks completely barefoot along the narrow lakeside trails with his Anglo-Saxon complexion and loyal yet retirement-ready field wear, Anton's aspect suggests one of two possible scenarios—that of an intrepid and seasoned wildlife researcher or last week's tourist finally exiting the forest after being lost for six days. Visibly more at home in the bush than he is in a conference hall, Anton is one of the precious few people who have been observing the state of the greater Gombe ecosystem (GGE) for the last 50 to 60 years, capturing the fluctuating trends of this remarkable place. It is with these people that the origins and the impacts of the Tacare approach lie.

"Because my specialty is wildlife behavior and ecology," Anton explains, "I think in these terms, trying to explain the relationship between the natural resources, the forest, the wildlife, and the villagers, and noting the effects the human community is having on the species around them. It's very clear for myself," he continues, "and for almost all people who've conducted long-term wildlife research, how much things are changing, and how much the creatures or their habitats are endangered or threatened by human activity closing in around them. Eventually, every researcher becomes a conservationist in the long run."

Anton raises an important point here. Although people generally consider wildlife research and conservation to be synonymous, the distinctions between the two disciplines lie in their respective goals. In its purest

Anthony Collins (*back row center*) was among Jane Goodall's students in Gombe in 1974. *The Jane Goodall Institute.*

form, wildlife research aims to fill the blank spaces in our knowledge by gathering new information for a given species. It begins with an observation that cannot be fully explained, therefore raising a question. Then researchers develop a hypothesis or multiple hypotheses and design methods to systematically test these through experimentation or observation. Once the data or results are gathered, they are analyzed. After a series of repeated tests, a question is answered, promoted to a theory, and published as new knowledge. In research science, this is the end of the line until the theory is challenged and the whole process starts again.

In science, a fact is only a fact if it can be explained by a theory; it is rarely absolute. This leaves all current knowledge open to better science in the future so we might refine or correct what we once believed to be true. Dr. Jane Goodall, for example, famously caused science to redefine the human descriptor of *toolmaker* when she observed wild chimps in Gombe modifying vegetation stems for task-specific behaviors like termite fishing, proving that humans are not the only creatures capable of toolmaking or innovation.

Conservation is an altogether different discipline and can in fact bring multiple disciplines under one canopy. These can include anthropology and cultural studies, ecology and biology, environmental and natural resource sciences, climate research, education, or even social and ethical activism. A conservationist should recognize the interconnectedness of all these fields. However, conservation as a formal science takes things one step further by combining experiment-driven (Western-style) research, traditional ecological knowledge, technology, and biotic and abiotic processes into programs aimed at rehabilitating or conserving the natural world through strategic development and operational, actionable objectives. These can appear in the form of conservation action plans (CAPs), land use or village land use plans (VLUPs), and participatory rural appraisals (PRAs), to name a few.

"As the years have gone by," Anton says, "I've become more and more involved in conservation initiatives. Gombe is a small park, and it became like an island of forest, surrounded by farmland with no chance for chimps to be reunited with other populations nearby. We were very concerned with trying to link these dispersed chimp populations together to ensure their [genetic] future. If we can connect these populations together by developing habitat corridors through the farmlands surrounding Gombe, then all number of species, the baboons and all the others, what they call biodiversity, would benefit." But simply connecting separate parcels of suitable habitat between two populations is not possible unless the focus moves from the needs of the animals to the needs of the people causing the habitat degradation in the first place.

Throughout much of Africa, this complex issue often includes unpredictable pressures brought about by war and famine. "One example to describe this complexity," Anton explains, "is that during this period [early 1970s], there were many immigrants coming from the Democratic Republic of the Congo (known then as Zaire) and especially from Burundi and settling in the villages closest to the park. At first, these people were not really living as part of the village community, and so were not subject to the village government." Although refugees have long been trickling into western Tanzania from neighboring nations, the civil conflict in Burundi between 1960 and 2000, specifically the genocides of 1972 and 1993, saw a flood of immigrants seeking refuge across their shared border with northwest Tanzania's Kigoma district. With Gombe's northern boundary just 12.5 miles (20 kilometers) from the border of Burundi, this wave of poor and desperate people stretched the local resources toward their limit.

As with most integration between immigrants and residents, there was a divide in terms of culture and even religion, which then translated to a spatial divide. "It was the locals," Anton continues, "who told them where they needed to settle, saying, 'Well, you can set yourselves up on the hill over there'"—he says this with a dismissive hand gesture—"and they were told to use the land between where their own villages were." This division unfortunately translates to more and more land being occupied by human settlements as cultural, social, and religious groups seek their own areas. As the floodwaters of immigrants began to rise, Gombe National Park became an island of green, carved out and encircled by the human-made ocean of bare hills and eroding farmland.

According to Anton, the challenges of having the Burundians placed in the middle of the forested areas meant that they were living in an unsustainable manner. "It was well known that they cut a lot of trees, hunted a lot of wildlife, and just harvested in an indiscriminate way," he says. "They weren't really under supervision by the village government and far less unified within the indigenous community. We realized this when we noticed people were coming in here selling charcoal, which was people's main cooking fuel at that time. We would ask these people, 'Where does this come from?' and they say, 'Oh, it comes from very far. We came from the hills, way away.' We then realized what had been happening for a long time. The Burundian immigrants had settled in villages right up against the park boundary behind us." Empathizing with their situation,

Anton says, "If you're a refugee or a fugitive and you come with nothing, if you can get ahold of a machete or an ax, a hoe to dig with, and a box of matches, you can start to make charcoal—but not without cutting down the trees."

Despite various other energy sources such as gas, oil, and hydroelectricity, Tanzania still relies largely on the use of firewood, usually processed into charcoal. In rural Tanzania, charcoal is almost exclusively used as an energy source, largely in a residential context. In fact, more than 90 percent of all local wood harvested is for cooking, and this alone constitutes the biggest contributing factor to deforestation today. Even in the nation's largest city, Dar es Salaam, where access to other forms of energy is available, nine out of ten households either rely on or prefer to use charcoal for cooking. With a population of approximately 60 million people as of late 2020, the pressure on forests for combustible fuels is at a critical level, because harvesting wood from trees is generating an open, soil-eroded landscape. With trees that were once drawing the subterranean water table toward the surface now gone, the underground portion of the hydro cycle is broken, leaving dry creeks and rivers in its wake.

Yet, despite the damage caused, charcoal production is often one of the few means impoverished people across rural Africa have of generating some type of income. The process for making charcoal consists of slow-burning wood either in a kiln made of clay bricks or in a type of underground oven, whereby a hole is excavated for the wood and then covered with branches, soil, and grass. Both methods allow for a slow and controlled amount of airflow, which reduces the rate of combustion. Instead of the wood burning away to ash, it becomes charred, hardened charcoal, which is then bagged and regionally distributed for household use.

Given the exponential population growth rate together with a dwindling supply of wood, the demand for this product often renders it unaffordable even for those producing it, forcing the poorest of remote community members to resort to daily harvesting of green, fresh-cut young trees and branches. The deforested areas are succeeded by grasses which, during the dry season, extend the spatial expanse of annual fires through an amplified fuel load. The increased frequency of this burning hinders both the regeneration of tree stumps and new seedling emergence and survivorship. With larger cut wood being designated for charcoal production and small sapling trees being used as fuel by those unable to afford charcoal, the entire life cycle of a tree is consumed at all stages of growth. Following rainfall, erosion then dismantles the landscape; in areas like the steep hillsides surrounding Gombe, landslides are an ever-present threat and have claimed whole sections of village land on several occasions.

"Wood harvesting," Anton says, "together with an expanding and shifting agricultural footprint, are land use practices for short-term gain, without long-term thought or understanding of the ultimate damage. Villages were losing reliable wood supplies, soil erosion was changing the land, and all kinds of options and resources were no longer available. There were new challenges to their lifestyle because they'd destroyed so many things in terms of vegetation, water supply, and soil. Within these areas, the land was now heavily degraded, and the animal species were removed to a great extent, and what remains are small pockets of forest here and there, a little bit of marshland, or some smaller vegetation."

Without alternatives or an understanding, it's common for such communities to completely dismantle what they once relied on. "Vegetation, soil, water, all these things can be preserved provided the villagers see a point in preserving them," Anton says, "but only if they collectively make that decision. It doesn't work if only some people see the point, because those that don't can just exploit them. It's the tragedy of the commons. 'Why should I refrain from cutting wood here if those guys are cutting wood over there?'" he says, illustrating this point. With this mindset, everything disappears fast.

"The main thing is to ensure that the people living there learn to value what they can preserve, and they see how preserving the woodland or the forest on the uphill side of a catchment does maintain a supply of wood for later use," Anton explains. "If they see that by allowing trees to grow, or planting trees and grasses along the contours of a hillside, they can stop erosion, that will preserve the integrity of the farm area and village land below. If locals know that protecting their natural resources can protect their farming resources in turn, then the cycle is positive and there are long-term prospects. It's really a question of both education—explaining how the cycles of water, nutrients, and climate work—and also getting them to see the clear benefit from their actions."

This point raised by Anton is a critical dimension of JGI's approach to developing community-led conservation. Sensitization helps the local people better understand the factors of cause and effect. This is where community mapping—a process of identifying relationships, needs, and resources and recording them in the form of a map—helps illustrate the variables involved and the results of either good or poor land management. For community mapping to be most effective, geographic information system (GIS) technologies are an invaluable asset. When, for instance, community members have access to a large satellite image of their village and nearby resources, they can immediately gain a sense of spatial proximity to other houses, water resources, forests, and farms, and correlate these to sites showing local disasters such as landslides.

In addition, imagery from years past helps show trends of change, both natural and anthropogenic. Through their own local knowledge, villagers can link these trends with changes in past behavior such as the expansion of a crop, the removal of a forest patch, scarring left by fire, or the slow but obvious disappearance of a once flowing and productive riverine. The power of community mapping is that the spoken word is not the beginning and the end of information sharing. With a physical map or maps, local people are shown, not told. They can see for themselves without the barrier of disbelief or mistrust. Visual examples of space and time are immediate, relevant to them, and within an easily interpreted context.

As described later in this book, the most significant aspect of community mapping is that by upgrading these group discussions to participatory mapping, people from the area begin to drive the ideas, using the spatial and temporal information from GIS to develop their own solutions. They are presented with a complete dataset via a series of maps and can exercise the autonomy to use this information to design their future of land use. This critical stage puts those individuals in the driver's seat, with JGI staff on standby to assist and, at times, facilitate community initiatives or provide access to start-up infrastructure.

However, simply showing a map still doesn't secure behavior change. All stakeholders need to understand and be willing to place some level of faith in a nongovernmental organization (NGO) that its suggestions will yield an improvement in their basic needs. "It is difficult, but this is why I think it has to be done on a community level," Anton explains from firsthand experience. "Their village government and the community have to agree and, for example, say, 'OK, this is the spring where the stream begins, we'll stop cutting trees here, and we'll preserve all the vegetation within a certain distance of that.' If they do that with enough key hot spots around their village, and neighboring villages do the same, then there is some hope for biodiversity to flourish as well."

When alternatives are offered or information is shared through education, science, or technology, these communities will have the opportunity to buy in and direct their own development. Indeed, not only do they need to collectively agree, but they must also drive the change from within their own realities. JGI can open the door, but it's the village members who have to walk through. To effectively engage, listen, understand, and at times facilitate, Tacare practitioners on the ground need to be ingrained into the region if they are to bring about lasting and positive change. This means conforming to the context of a place, without allowing the lens of perception to become clouded by their own background or country of origin. Long-term commitment is a crucial ingredient of Jane's holistic philosophy—long-term commitment not only from the NGO but also from staff.

After all, it is with people that relationships are built, not with logos on fleet-issued vehicles. If you are to help local people develop strategies to drive their own sustainable futures, it takes trust, and trust comes with positive experiences over time. So, is there a point at which an advanced education and a large backpack full of the latest technology is more valuable than the perspective of an impoverished local, or a committed, field-based staff member? No, there is not. A highly educated, well-resourced scientist walking into an isolated African village to assume the role of an authority figure is on a fool's errand.

There is need for both, however, and a marriage of emerging technologies with current scientific understanding provides support for remote conservation field sites. But it is just that, support. The true and meaningful change needs to be more organic than just facilitation from external parties. An unintentional yet all too common limitation of NGO initiatives is that they arrive with a set of deliverables, a matrix of key performance indicators, and a few years of funding to somehow achieve lasting change. Afterward, the NGO withdraws with a list of services provided and enough color photographs to return to headquarters and write a report. This is understandable, since the projects are usually donor-driven, and virtually every donor wants to see their contribution manifest as a successfully completed project. For this reason, educating donors on the realities of sustainable community-led conservation is in itself a large undertaking.

Despite the best of intentions, this common NGO approach is unassimilable. True change comes from within people's realities—their context—and is only sustainable if adopted and driven by the local community. Even though this may not always begin smoothly, if an NGO truly puts down roots and integrates itself into the realities of the people, then relationships of trust develop, and this gives rise to a collective growth of organic and positive progress.

Today, this insight continues to inform Tacare initiatives across a broad spectrum of JGI's Africa Programs. Although the approach has been shaped by many lessons learned, it's a practice that's still evolving and will no doubt be refined through various challenges in the future. In fact, Tacare wasn't always this holistic in its approach, this inclusive of all the social and environmental variables. An ongoing process of trial and error forced one key attribute, adaptability, to become a keystone of JGI's approach to community-led conservation.

Driven by the urgency to address the systematic deforestation of the greater Gombe ecosystem, JGI staff would soon learn that the social and economic issues affecting rural communities run far deeper than the root stumps left clinging to the now barren hillscape. In the following chapters, Jumanne lights the way for a younger generation as he carries the torch for what will become JGI's most far-reaching initiative, while Anton and other program founders share insights from the official inception of the TACARE program, illustrating how a focus on the environment and its dwindling vegetation was in fact overlooking the region's most immediate humanitarian concerns.

Chapter 2

Paradigms and problems

Jumanne Kikwale moves back to Kigoma to teach the next generation about trees

Anthony Collins recalls Tacare's earliest steps—and missteps

After returning to Gombe in 1974, Jumanne spent a year working on field research with Anton. But just as the distant contact calls of a chimp group help guide outlying individuals to return from foraging, the lure of education called Jumanne out of Gombe for almost two decades. "I attended a teaching college in Mwanza," he says, "and after my training was complete, I stayed for more than 17 years working as a teacher. In December of 1991, I came back to work for the Jane Goodall Institute, first as an administrator at the Gombe National Park, and then Jane introduced me to Roots & Shoots."

Celebrating the milestone of 30 years of research in Gombe, JGI's education program began unofficially earlier that year, when Jane gave youth outreach talks at schools in Kigoma and Dar es Salaam, Tanzania. Aimed at educating and engendering compassion for animals and the environment, the program evoked such a positive response from 12 individual students in Dar es Salaam that they formed peer action groups, who then rallied to voice their concern for issues of animal welfare and environmental degradation. The response led to the formalization of what

Jane (*back row, third from left*) with the original 12 Roots & Shoots students in Tanzania in 1991. The group has grown into a worldwide network that empowers young people as they address community and global challenges they are passionate about. *The Jane Goodall Institute.*

went on to become the Roots & Shoots program. Who better to carry this fledgling program from Dar es Salaam back to the Kigoma district than Jumanne—a seasoned schoolteacher, already woven into the fabric of JGI and, more importantly, a local to the greater Gombe region?

Human history has shown that the biggest paradigm shifts happen during a transition between one or more generations. Though humans would like to believe that it's their superior intellect causing them to question or challenge what their predecessors did, much of the drive to change is likely influenced by the fundamental processes of biology.

At a base level, organisms are programmed to strive for opportunity. Most plants and animals show innate traits of challenging limitations passed down to them, to dive deeper, fly faster, push their stems taller, generate broader leaves, coordinate and cooperate, problem solve, and even manipulate their physical world. Driving these innate mechanisms are three basic needs—to secure space, nutrition, and reproductive success. But the more an organism can advance its adaptive strategies, the more it can thrive.

Most biological scientists refer to this as evolution—the spark of life that engineers us to improve with each successive generation. And this is the spark behind Jane's vision for the Roots & Shoots program. By engaging youth across all age groups, conservation educators can plant seeds of positive change. Those seeds then grow into tomorrow's decision-makers, the people with the ability to act and refine what was done before them. This is how paradigms shift, and this is why Jumanne was perfectly positioned to influence the next generation of conservation leaders in western Tanzania.

"I introduced this program in the villages around Gombe National Park," Jumanne explains. "I started with five schools—Mwamgongo, Mtanga, Mlati, Bubango, and Chankele primary schools. I remember when I was young, much of what I learned was about keeping the forest. How if we need to cut old trees, we must also plant young ones to replace them for the animals, but also for our own future use." After learning that people shouldn't harvest trees without creating a cycle for the years to come, his thinking began to shift. Later, when he went to introduce these ideas to the community teachers, they accepted the concept directly, agreeing to bring environmental education into their schools.

Although Jumanne mentions this rather casually, asking teachers to insert a completely new subject into their school curriculum isn't always a simple or successful undertaking. Such engagements require physically traveling to the remote villages, meeting in person, exchanging ideas, and holding lengthy discussions and explanations of intent.

Jumanne was well situated to approach this challenge. By selecting and supporting the right team members, established NGOs have the advantage of recruiting key local staff such as Jumanne, who has the added credibility of being a former schoolteacher himself. "Next was to put these ideas into the minds of the young people," he continues, "to

Jane Goodall (*standing at left*) and Jumanne Kikwale with villagers near Kigoma, Tanzania. *The Jane Goodall Institute, Chase Pickering.*

learn and understand their environment, and the benefits of conserving what's around them." Again, as simple as it may sound, this step highlights an aspect of community development that can be difficult to influence—that of social and environmental imprinting.

There is inherent difficulty in changing a social or environmental imprint, especially if the change threatens to increase hardship from that person's perspective. Because an imprint governs how children will grow to interact with the world around them, when an NGO arrives in a developing region to impose a completely new way of existing, what little change is achieved typically erodes once the NGO withdraws. Not only is this practice unsustainable, it also hinders future NGO support by leaving behind a trail of unsuccessful interventions, making it difficult for other outsiders to achieve trusting relationships in the years ahead.

From the perspective of children in rural Africa, the principles of environmental conservation may challenge much of what they know, and if not approached carefully, verge on threatening what little they have for surviving daily life. From a child's perspective, learning that 2 + 2 = 4 does not mean you'll go hungry, but learning that you shouldn't harvest firewood from the national park for cooking does. Once a child reaches adulthood, the social imprint is ingrained, which can mean that engaging with older generations on topics of social change becomes more complex.

The legacy of the Tacare approach is that change comes from within, and it comes slowly. It is change that is driven by the members of the community after years of work have gone into engaging, listening, and understanding the realities of people's lives. Only then can alternatives to unsustainable practices be identified and agreed on. Jumanne's task of changing how children view and use their surrounding resources is a key element of this approach, and Roots & Shoots provides this platform.

"I think it's fantastic," he says, full of enthusiasm, "when educating them about this real-world subject, you can see the children beginning to understand. Even now when you go into the schools and mention conservation and the environment, they're aware. They know the value of the forest, they know the value of the animals, and they know the value of the people. It is working very well in their mind, and we hope that in the future there will be no need to educate. They will know from the beginning, from their own parents while growing up."

As an example of Roots & Shoots' work in the community, Jumanne says, "We have nurseries in the schools with different species of tree, and when these are ready to be transplanted to a farm, the children take the plants from the nursery to their homes. Their fathers used to ask them to bring more seedlings, so they could plant more. Even before the generation changes, today's adults are learning about the environment through the program." Although some alternatives can be provided, such as the trees from the school nursery, Roots & Shoots is not just about providing the answers; its aim is to illustrate the connectedness between humans and the natural world so that children learn to question the status quo.

"As we educated the young people and their views began to shift,"

Jumanne continues, "school students took this education to their parents at home." Understandably, being taught by their own children that they had played a major role in depleting village resources such as trees and accessible water wasn't always well received by the adults. Jumanne recalls an incident from the early days of implementing Roots & Shoots: "We had a tree nursery at Mlati primary school for growing and distributing saplings and planted the extra trees around the school. Unfortunately, sometimes through the night, people would go in and cut them down.

"The resistance came from those who had goats and wanted to let them graze through the area we had planted our trees. We tried to figure out who was doing it but couldn't, so we went to the village government and reported it. They caught one of them and unfortunately, I have to say, the one they caught was a teacher." Jumanne says this with disappointment, almost taking it personally on behalf of the teaching profession. "It was dealt with by the village government and luckily, it didn't happen again. In response to those who wanted land for grazing, we then held a community meeting in front of the village government to address these issues. We publicly formalized that certain areas were designated for trees, and others for grazing. It was successful and didn't happen again. This is how we tried to overcome any problems. The change was driven from within the local government and the people."

During this period, Anton, too, was involved in Roots & Shoots. He'd been active in the program since its inception in 1991, going around the villages closest to the park with Jumanne, but at the time he was also busy working with others from JGI on a grant application. Eventually, the application reached the European Union, which in 1994 provided a start-up grant for a new program—a project that began as an acronym and evolved into an approach. "When we received the grant, this very wonderful project called TACARE—Lake Tanganyika Catchment Reforestation and Education program—was born," Anton says with pride. "From then on, it was like our baby."

Tacare started slowly, from five villages, to 12, then 25, until some

Dr. Anthony Collins with baboons on the beach at Gombe National Park.
The Jane Goodall Institute, Bill Wallauer.

iteration of it has reached 104 villages in Tanzania today. Back then, in the beginning, Anton says, "one of the strong points was the personal relationships the team had with the villagers. Initially, it was tree planting that was the major mission, to protect Gombe, rehabilitate the bare hills, and mitigate the damage already done to the vegetation. Trust built up gradually and firmly by these repeat visits back and forth, explaining why we were doing this. Secondly, quite early on it became clear that no matter how fast you plant trees, and no matter how well you secure their future against getting eaten, trampled, or burned, we couldn't keep pace with the tree cutting because of the people's demand for firewood. If people are struggling, there's no reason for them to respect all this nice tree planting because at the end of the day, they needed fuel for cooking.

"Also, many people were suspicious. They said, 'Yes, but we know what you're doing. If you get us to plant the trees, as soon as there's more decent forest here and the animals start to come out, then the government will move us out. The national park will be extended. We will lose our land, we lose our livelihoods, we will lose everything we have.' Of course, that's a nasty fear to have to overcome. I think the staff overcame it simply by being very straightforward, trustworthy, honest, by continuing to go back and go back, and also I think they were slightly helped by not having too much money at the beginning because the project grew slowly." With this approach, JGI staff built trust because the villagers could see which way they were headed and what they were trying to do with greater clarity than if Tacare had been launched on a larger scale.

"What I rather respect about the way Tacare has been run," Anton continues, "is how we dealt with opposition, and there's been a number of villages that were indeed oppositional. JGI staff would say, 'Well, if you don't want this, then we're leaving. If you object to this, you won't have anything. We're going,' which is a strong response. When they would leave, it was always a sad moment, very sad to have to break it off. The usual outcome was after re-elections came around and a new village government came in, Tacare was invited back. The fact that the villagers actually invite them back and really want them is a nice justification that there is some real help going to the village and the locals can see that.

"The second stage was when the team went in and sat with the villagers and asked them, 'What are the key things? What matters to you most in your life? What are the main challenges you have in your life?' After people were given this platform to explain their concerns to us, trust took a major leap forward because they'd say, 'These people are actually coming to help, they care about our welfare. It's not just tree planting.' I think that was a wonderful thing, and we should've done it this way from the start."

In one season, the JGI team broke up into agriculture, habitat health, and community development branches. "Working on these separate aspects meant we got money from different donors for each," says Anton. "Then the local people really saw how we had redirected our focus from just tree planting to issues relevant to them. They began to think, 'These people are with us, they're not like an NGO any longer.'" In Anton's view, the program built its strength by starting slowly and growing gradually, adapting in response to the locals' feedback, and being straightforward and consistent, developing personal relationships with the villagers.

Anton notes that Tacare staff also developed an open dialogue with the local government, including the village councils. Through this engagement, JGI became more aware of the environmental threats posed by the floating population of immigrants and refugees from Burundi. "The whole area, at least on the south end of the park, was almost stripped naked by charcoal burning and conversion to agriculture" he says. "Even we didn't notice this for a while because we were in the park concentrating on the wildlife aspect. It was a while before we realized the very serious effect that these people who are not part of the village were having on the village land."

Here, Anton pauses to reflect on lessons learned. Tacare evolved through trial and error, and, in retrospect, he sees that their initial approach, while well intentioned, wasn't always as effective as they had hoped. "Knowing what we know now," he says, "I would have switched from reforestation earlier because although the initial program was focused on planting trees due to them being overharvested, erosion occurring, and the habitat falling to pieces, it wasn't an immediate issue from the villagers' perspective. Apart from this not being the most relevant issue the local people had to deal with, another realization that came was that tree planting is labor intensive, it costs quite a bit, and even if you do plant a tree, you can't just leave it there, you have to make sure it survives the pressures of the landscape for long enough to grow above the height of grazing animals like goats and be large enough to be resilient against fire and so on. Many planted trees never make it, so there's a rate of loss. Most of the ones that thrived were planted not on the hillside but in the village itself because vegetation around the houses received better care and wasn't confronted with overgrazing or fire. Those trees we planted further out there, well, they didn't get a chance. The fact is that tree cutting was a major threat, but tree regeneration is a very specific and natural process. It's not easy to replicate on a large scale."

Anton points out that if a forest is destroyed, the land is very forgiving. If you protect the land from fire and you keep cutting at a low level, many plants can just regenerate. The seeds that are in the earth can regenerate and, provided there's no fire for three or four years, they can grow above the height of the fire. Fire-resistant trees, of which there are many in the region, can then thrive and the land can recover. But if the fires come through every year, the area will remain a grassland with little trees that look like bushes—the stumps of large trees trying to re-shoot. Then a fire comes through and takes it all off again.

If you can keep the area as a fire exclusion zone, Anton explains, the trees come back rather quickly. But, for the initial TACARE project, that realization didn't come till after a lot of investment in tree planting. Tree planting has its place, he says, and has many benefits, but in terms of the big, broad hillsides around Gombe, what was needed was fire control and a reduction in cutting. "If we'd have done that earlier, we could be way ahead of where we are now, as shown by satellite imagery and so forth," Anton admits. "Anyway, things are going very well since that point, but it would have been better to start that earlier."

Anton recalls some other initiatives that didn't go very far. For example, he says, in the village health program, money was raised to buy toilets for schools in some villages and for the development of a publicly accessible village toilet. "Of course, the people were happy to have [the toilets], they were nice and well-received, but once the project finished, it became clear community members were not looking after them. The villagers often had the attitude of 'When are you going to come and repair your toilets?' Like it was up to us to go, rather than that ownership being taken up by themselves."

In general, Anton believes, transfer of ownership is a challenge. The toilets were, he says, "a rather crass example, but the issue arose in other aspects of the project as well. You want to improve their lives so that when you go away, things are better. You've left them better than you've found them. Some things had to be pushed a little further or explained a little

An aerial view of the northern border of Gombe National Park on October 31, 2000. Mwangongo Village (*on the left*) shows the impact that human settlement has on forests, watersheds, and the chimpanzees' ability to disperse northward to Burundi. *The Jane Goodall Institute, John Maclachlan.*

further in that way. One other stumbling block is if you work with the local government, the village government, through elections, they will change every few years. You might bond very well with one village government, then a new governing team comes in and maybe has a different agenda, perhaps even a selfish or nepotistic agenda. There's been a few cases where we've had problems with certain villages due to a change in governance."

For the sake of capacity building, aid programs or NGO programs may plan to eventually withdraw, which, Anton says, is the ideal. But he's not sure whether the ideal is always attained. "There are many cases of them withdrawing and everything falling to pieces after they've gone," he notes. "The essential thing is to extend awareness and vigilance in the community to think long term for the benefit of their people and the natural resources they depend on. If you can get that critical number of the population valuing and thinking long term, and secondly, feeling that their lives have actually benefited from whatever interventions the NGOs made, then I think you can withdraw."

As an example, Anton mentions the coffee program that was run by

Tacare. "It seems those coffee farmers are doing extraordinarily well compared with the lifestyle that JGI found when they went there," he says. "They're living in much better houses, children are being put through school, people are traveling more, they're starting their own businesses—a real quantum increase in the quality of life. If we withdraw from that, it's fine. It's theirs. It's not ours. As much as possible, village-specific projects need to be run by the local community. They can take pride in what they're doing and that leads to ownership."

Reflecting on other challenges faced when trying to balance environmental and humanitarian needs in rural Africa, Anton mentions climate change as an obvious example, but he says, "the thing that worries me most is simply population increase." No matter how good an NGO's recommendations for managing the land sustainably, "an exponential population increase in a village means they need more farm area, so there's going to be some conflict between what they want to do and what an NGO's recommendations are. To that extent, it may be necessary for NGO programs or aid programs to at least keep one foot involved in a community, to continue to advise and try to be aware and keep ahead of what is likely to happen."

For Anton, the key to long-term success is that the communities themselves must see a value in preserving whatever has been implemented. Crucially, if the program doesn't include any income-generating avenues, such as honey production, mushrooms, medicinal plants, or controlled access to firewood and timber, the local people don't benefit; nor will they appreciate the change. The good news, Anton says, is that "within the national park, the forest is regenerating very well." Outside these protected areas, there is still an urgent need for threat-mitigating initiatives, and it is in these areas that Tacare carries out the long, slow process of putting down roots in a community, understanding its needs, and seeing green shoots of change begin to sprout.

Chapter 3

1994: Understanding deforestation

George Strunden and the genesis of Tacare

"At the time I met Jane, I was working for a German government agency in western Tanzania," recalls George Strunden, an agricultural scientist with many years of experience in regional Africa. "When JGI received funding from the European Union to start the TACARE project, we met to discuss a role at JGI and then I started in 1994."

With his professorial manner, George sounds matter-of-fact about this moment, and yet he is describing the genesis of the Tacare approach as it exists today. The very first official TACARE project was initiated and led by George, who had spent four years in a development context in the Kigoma area outside Gombe National Park, where Tacare began. George started as a project coordinator and, in the next phase, became project director. Regarded as the father of the project and a key character in its evolving legacy, George has also played a significant role in mentoring JGI team members either directly or through his example. A great many of JGI's former and current staff members still draw on his leadership, while his pragmatic approach to complex social and environmental issues is one that continues to resonate through JGI's Africa Programs today.

George explains that Tacare, initially designed as a reforestation project along the shores of Lake Tanganyika, went through two phases funded by the European Union and then by other European foundations. As it attracted other donors, including the United Nations Development Programme (UNDP) and the United Nations International Children's Emergency Fund (UNICEF), more components were added to it. But when it started, George says, "we named it TACARE as an abbreviation, which was also pronounced *take care.*"

Echoing Anton's account, George describes how Tacare staff came to realize that regeneration of the natural vegetation was more effective than tree planting. "We had some examples in Kitwe where we observed that the regeneration capabilities of the native Miombo woodland biome is quite strong," he says, "and that to reforest the hillsides around Gombe National Park and in the south would be more effective if left as management areas rather than planting trees because of the frequent fires."

Although fire regimes are a typical function of any ecosystem with dry grasslands, the open and dense grass replacing the larger woodland tree and shrub species that had been cleared on the hillsides created unnaturally frequent and intense burns. The fuel-laden open hillsides created the perfect conditions for intense fire regimes, lowering the survival rate of young, freshly planted and ill-equipped tree saplings. "The Miombo vegetation is tolerant to fires, but with regard to planting, if it was not Miombo, it would not survive," George explains. "To plant the Miombo trees is difficult, and it's also complicated to raise them in nurseries, so we pretty soon realized that there's more a landscape management effort that is required than a reforestation effort."

Like Anton, George now sees these early tree-planting efforts as well-intentioned but misguided: "The vegetation succession in these situations is better by itself if given time to recover. The people who designed the initial project before we all started thought they could plant trees to reforest these hills, and that didn't work out. That was not the right approach." Even so, Tacare's presence in the local communities had a positive outcome. "What worked well, I think, was our initial engagement with the population. We spent a lot of time in the villages. There was no real land-use planning happening then. The decision to move away from tree planting toward management of these areas was probably the biggest step we took.

"We went with a team of people and camped in the villages. We spent a few days and nights per village and had a series of meetings with the community members. The atmosphere in these villages toward Gombe National Park and JGI was rather negative because they were worried that the effort would be to expand the park, so we had to put a lot of effort in building trust and then convincing people that this was not part of the plan. Then when we agreed upon the first development activities, like spring and riverine protection and similar kinds of activities, we slowly managed to earn some recognition as trustworthy advisers."

George Strunden. *The Jane Goodall Institute, Victor de la Torre Sans.*

As part of this relationship-building, George's team applied certain techniques with these communities that were highly successful. One was the LePSA method of introducing the project activities. LePSA stands for Learner-centered (Le) training, which involves three phases: Problem-posing (P), Self-discovery (S), and Action (A). As George explains, "That is the technique where you describe a situation, and then encourage a discussion within the group you are talking to, and that worked very well because it was inclusive, encouraged participation, and showed we were there to listen more than talk. We did that village after village. We had several discussion starters; we had an image, a painting with deforestation

Jane Goodall (*left*) and George Strunden inspect a Tacare tree nursery in Kigoma, Tanzania. *Michael Neugebauer.*

and erosion." But, George says, the most effective opener was explaining that they had just come from a certain village at the Burundi border, which everyone knew had no trees left. By then, these villages were also struggling with the impacts of deforestation.

"They had erosion problems and no access to firewood. We described that situation, and then we opened with a question, saying, 'What about in this village, is that a situation you face here as well?' Then others in the group said, 'Well, actually, we are having a similar situation,' and then we could step back, and they started discussing it among themselves. Then we picked up on that and said, 'Well, if that is the situation here as well, what do you think could be done?' Then they suggested things and we just guided the discussion a little bit, but basically the conclusions, what needs to be done, came out of these meetings. That worked quite well, and we continuously applied that approach.

"They all had these barren, eroded landscapes, so the discussions were always around, 'What can we do to get our trees back and to resolve these problems?' We didn't offer solutions as much as we prompted the questions, which were then, 'What can be done?' and 'Who could go and do it?' We didn't present ourselves as the solution to all the problems, so we kept it in their realm."

George acknowledges that, "in some villages, the atmosphere was a bit tense," but, he says, "we had time, and we could keep things very general for a while. They didn't have many visitors in those days, so we were somehow guests. Since we didn't have to cover it all in one meeting, if it got too tense, we could talk about other things and then start the next day again. I remember Kiziba was one of the villages where the situation was quite tense, but then it changed after slow discussion.

"I don't think we met any resistance as outsiders because some of us were white, but more so because our discussions were in Kiswahili, not in Kiha, which is the local language. We were all outsiders in a certain way, we were not locals. Even people from other parts of Tanzania, the fact they were speaking Kiswahili meant that they too were all considered to be outsiders."

When Jane Goodall began working in 1960, villagers were still living in parts of Gombe. Then, in 1968, the Tanzania National Parks (TANAPA) stepped in and Gombe was declared a national park. There had always been some restrictions because the area had been a game reserve since 1943, but many locals believed that Gombe's designation as a national park was somehow linked to Jane's research there. In fact, George points out, all the law enforcement and the displacement of people happened through TANAPA—it was a government intervention.

"The biggest forest destroyer in that area was conversion to agriculture. In the Ujamaa period [a period during the 1960s and 1970s when Tanzania implemented a collectivist policy based on cooperative agriculture and 'villagization'], people were concentrated into villages, and all these lakeshore villages, they grew in size and population. People were basically just moved into these villages from a more dispersed lifestyle; plus you had population growth. There were suddenly more people in these villages than traditionally in the past. They all engaged in agriculture because they were not supplied with any food. So after they were moved, they had to grow food pretty fast. The dispersed families converted these Miombo woodlands into fields, and many of them are not suitable for agriculture. The soil is not fertile, and they're very steep, so they had one, two harvests, and then the topsoil was gone, then they gave up. Then the vegetation slowly recovered, but these little trees were harvested for firewood, fish smoking, and construction wood, and so on."

In discussions with the villagers, George's group explained that those steep areas were not suitable for agriculture, and that if the villagers harvested those trees too early, the firewood yield would be low. They used a couple of scenarios as examples. In the Kitwe forest area, Tacare had already regenerated some forest, which they could use to demonstrate to the people how these forests would look after a few years. Local individuals also shared their own experiences where they had seen how these forests were coming back. The forest regeneration in the Kitwe area took place after Tacare helped improve silvicultural practices, such as the slashing of bushes and grasses and the thinning of undesirable coppices. Tacare staff also implemented improved wildfire management activities, such as fire lines and fire breaks and timely prescribed burning. Similar improved forest management practices were adopted in other forest areas, which multiplied the effects of JGI's conservation efforts. Overall, in the JGI conservation program areas, several of the once-degraded forests have successfully regenerated on village lands and are now managed as village-owned forest reserves.

George admits that, at first, progress was slow. "Although there was and has been progress, behavioral change in that area is a very difficult thing to achieve. It took quite a while before we could observe a change in attitude. Still, from a hostile environment towards a collaborative environment, we could observe that change after the first few years. When we started, there were discussions between the locals and the Tanzanian government to give Gombe National Park back to the communities. And when TANAPA maintained the park's protective status, there was a real negative attitude toward the park for a while afterward, and us by extension, which we worked hard to overcome with assistance initiatives like land-use planning.

"We engaged some 10 lakeside villages and 10 highland villages, who adopted our facilitation of individual land-use plans, and we were able to influence them in a way that their forested areas created this interconnected habitat corridor. We managed to have 20 villages pull on the same string. When it first became clear that this methodology was working, it was quite a large feeling of success. Then over time that result could be observed from space using satellite images. We could see a positive change in vegetation recovering in that area, and it was the result of five to ten years of effort to get there." Despite the positive change, George admits they could have done better. "We didn't start the land-use planning effort right in the beginning. If we would have engaged in that immediately, we would have achieved some change earlier. That was also a process of thinking and learning by trial and error, by being adaptable and responsive in our approach."

When they started the land-planning effort, George says, one issue they faced was land title and ownership. In the areas in question, the upper hillsides that the chimps were still traversing, there had never been any demarcation between the villages. "The boundaries on the lakeshore between village A and B were very well defined because that's where people were moving a lot. But when it came to the eastern border of the lakeshore villages, the boundary between one village and the highland villages was not defined and required a committee from the lakeshore village and a committee from the highland village to physically determine that boundary, which was one of our earliest spatial or mapping efforts. It wasn't until 2000 when Dr. Lilian Pintea joined the JGI team that we started to benefit from the use of geospatial technology."

At the time, Lilian was a PhD student in the conservation biology program at the University of Minnesota, working with Dr. Anne Pusey at the Jane Goodall Institute Center for Primate Studies to apply GIS and remote sensing to chimpanzee behavior research and conservation in Gombe. He first mapped the forest cover change inside and outside Gombe between 1972 and 1999, using satellite images from the Landsat program acquired by NASA and USGS. Then he introduced high-resolution one-meter satellite images from IKONOS when it became available over Gombe in 2001, which allowed local communities, Tacare staff, and other partners to see every tree, house, or foot path for themselves. Lilian worked closely with the Tacare team to bring all this information together in a GIS using Esri technologies and to develop maps that were useful for decision-making. These maps and analyses were used over the years to inform the evolution of the Tacare approach.

The remote sensing and GIS technology was an enormous novelty for JGI in terms of better understanding the landscape. "We had some old aerial photos," George recalls, "and 1:200,000 maps with contour lines, so we could see a little bit of what was happening up there. One tourist camp operator flew with us over these areas so that we could physically take oblique aerial photos, although that provided limited spatial insight. The remote sensing imagery gives you great information on where the vegetation was changing over time, but it doesn't always give you an insight into why it's changing, and it doesn't give you any strategy on how you can influence people to alter the way they do things there. That still lies in the relationships you have and the discussions you have with people and how you engage with them to effect change."

In general, George says, older men and older women are influential in the local communities, but ultimately it's a question of finding individuals who are open to change and collaboration. "The land committee is usually composed of representatives of the families who have been in the area longer than others. They can be old men or old women or both, but it's usually these elderly groups that have the answers. They know where

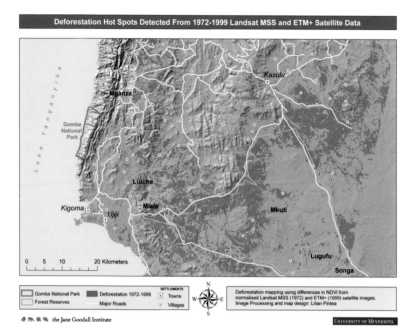

Deforestation Hot Spots Detected From 1972-1999 Landsat MSS and ETM+ Satellite Data

An example of an early map developed by Lilian and used by Tacare showing the extent of forest and woodland loss (shown in dark orange) between 1972 and 1999 as detected in Landsat MSS and ETM+ satellite images. Lake Tanganyika is shown to the far west, Gombe National Park is on the coast, and forest reserves, towns, and villages are also included. *The Jane Goodall Institute, Lilian Pintea.*

the boundary markers are, and these people are very influential because they also allocate land use to family members and outsiders. Without that group, you cannot achieve any change, so the land committee is a very important group. Then more educated people are also very important: the teachers, the government representatives, and civil servants who have been working somewhere and then come back to the village. These people are often influential and important to engage with, listen to, and understand. But I can't say that there's a specific demographic who was more willing to change than another. It's more about individuals.

"A big part of the work in effecting behavioral change was that you

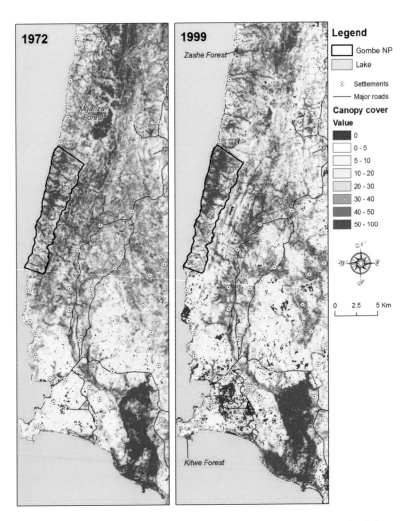

Maps of tree canopy cover change in and around Gombe. By 1999 (*right*), tree cover had increased inside Gombe but was showing significant loss or conversion to exotic species, such as oil palm, outside Gombe. These maps were used by Tacare staff to adapt their conservation strategies. *The Jane Goodall Institute, Lilian Pintea.*

had to understand and show concern for their priorities and their needs. It was a combination of identifying their perceived priorities and what we could get funding for. We didn't have a big unrestricted pot of funding, but we could build into project proposals certain components that we knew were important to the people. When you look at the social economics of these areas, agriculture is the backbone, so that is always close to the top of the list in terms of what they think is important. This was closely followed by their concerns of health care, social infrastructure, and water. Water was always very important to people." George notes that water has an enormous impact on the quality of life in the villages—waterborne diseases, the time people spend to collect water, and the quality of water are all issues that locals want to address. At the time, Tacare's choice of community development projects depended on a mixture of pragmatism—projects they could get funding for—and community priorities.

Here, George highlights two key elements of the Tacare approach: pragmatism and adaptability. "In fact, adaptability is critical. When you pursue conservation objectives, you need to have different tools at your disposal and be prepared to abandon the original project plan if you identify its flaws. Specific to conservation, the most effective tool is of course to establish a protected area by removing people—not allowing any activities or exploitation—which is great but limiting. The amount of land you can protect in that way is small and not enough to recover a declining species such as chimpanzees. You may extend these protected areas with surrounding buffer zones, but in western Tanzania, for example, regarding suitable chimp habitat, around 90 percent of the chimpanzee range is not in the protected areas. There are two national parks with chimps, Gombe and Mahale Mountains, and that's it. So, 90 percent of the chimpanzee habitats basically are on village and district land.

"For these areas, I think something like a Tacare approach is effective, but it shouldn't be the only conservation tool. What I would tell future Tacare or conservation and community development practitioners is what Lilian [Pintea] helped us a lot in figuring out. You need to be clear on what the conservation objectives are, the most important threats, and the priority actions and strategies. What are you trying to achieve? You need science and then science tools to determine those [objectives, threats, and strategies], and they have to be realistic. They're not 'Let's try to protect everything,' but 'Let's find out what we want to achieve and prioritize.' Then usually it is around habitat connectivity, and it's also preventing fragmentation and keeping the [wildlife] populations by having enough genetic exchange with other populations. That is probably the thing to achieve. However, in answering these, you need to do so with the local human population in mind, so when the inevitable time comes to engage with communities, they're factored into the broader picture."

According to George, conservation practitioners don't always make a point of showing understanding and respect for local communities—taking people seriously, trying to understand their point of view, and understanding their socioeconomic context. "What is their livelihood? What are their aspirations? You must be truly clear on what these are. If you are not clear on those, you will not be able to engage with them. I think one especially important thing, which we spend a lot of time doing with them in the Tacare approach, is to plan together. If you impose plans, you will not get anywhere. The people have to feel that this is something they came up with and they decided to do."

Through the Tacare approach of listening, understanding, and adapting, it soon became clear, George says, that "alternative livelihoods are what reduce the unsustainable use of natural resources." And so, developing alternative livelihoods became an important element of Tacare's work with the communities. From a focus on reforestation to sustainable agricultural projects, Tacare was evolving—as Aristides Kashula explains in the next chapter.

Chapter 4

1994: The forester

Aristides Kashula sees both the forests and the trees

"Local ecological knowledge is very important," Mzee Aristides Aloys Kashula says. "I will never accept the idea that local ecological knowledge is not important enough to be factored into community development, because I've seen its importance. There's a lot of typical research that has been done, and most of this is not being used for anything. It's like that knowledge is just sleeping somewhere in the cupboards."

Trained as a forester, Kashula (as he is respectfully referred to) has been involved with Tacare since its earliest days, and his pride and enthusiasm are evident as he describes his decades of experience with the project, as well as the local ecological knowledge he has gained, and applied, along the way. His association with George Strunden goes back to the early 1990s, predating the genesis of Tacare. Between 1991 and 1994, Kashula worked as a forestry technical adviser to George, who was coordinating a soil erosion project based in Kigoma, Tanzania. "We worked there together for almost three years before George then moved to JGI," Kashula recalls. "Soon afterward, I was seconded from the government by George's invitation to support the initiation of the TACARE project in 1994."

After working for a couple of years on the TACARE project, Kashula was employed by JGI on a permanent basis—and has been there ever

Aristides Kashula. *The Jane Goodall Institute.*

since. Over time, he's held various positions, such as head of forests and agriculture, and is currently the forest officer for the latest initiative, the Landscape Conservation in Western Tanzania project. "In total, I have been with JGI for almost 25 years," Kashula says with pride.

Reflecting on the early days of Tacare, he says, "I remember when we began way back in November 1994, we spent about five months trying to familiarize ourselves with the communities, the district local government administration, their environmental situation, and the capacities of their local government. It was a preparatory approach. Very quickly we saw that better-informed agricultural practices could play a role, not just in terms of food and nutrition, but also socially." The communities around Lake Tanganyika had previously relied heavily on fishing, which in turn had depleted the resources of the lake. So, Kashula says, "We had to teach them how to grow vegetables."

Crucial to the success of these agricultural projects was the participation of women. "One of our agriculturists at the time became focused on including the local women. We had more women attend the meetings as a result, but when we began dividing vegetable seeds for everyone to use in their agricultural activities, the women all left. We realized this was because it was only men dividing and handing out the seeds. There was an obvious need to incorporate women into the community development positions so they could mobilize other women in the village to participate during the planning meetings."

A woman trained in fisheries was nominated by the local district

An outreach and sensitization meeting in which women are sought-after participants, Tchimpounga Reserve, Republic of the Congo. *The Jane Goodall Institute, Fernando Turmo.*

government to fill this role, Kashula says, and to engage on behalf of the important responsibilities women have in the society. "With changes such as this, you need to be passionate with the approach in order to make it happen. Because I was young at the time, just married with a single daughter, being involved with the altered social system earlier in life meant I remained committed to making sure this change continued. Then women became participants, which was important because they're the ones who are really affected by whichever system is established or destabilized.

"After all, they are the ones who are caring for the kids, they are the ones who are caring for the household. They are also affected by the destruction of the environment. If water is not there, women are more affected because they need to gather water from much further away, and when firewood for cooking is hard to find, it's the same—they're impacted because it consumes more of their time. By virtue of our culture, women are always trying to stay away from whatever discussions or decisions are taking place. But if they did attend, when asked a question, they would just say yes, not really meaning it or truly being invested in the discussion. So, if policies are not favoring them, they are not internally motivated to be involved."

The next step was to identify more community development activities, such as alternative income opportunities and erosion mitigation, and to hold a conservation planning workshop. The workshop included components on water, savings and credit cooperatives, water infrastructure, and infrastructure support related to health. "We also realized we needed to give focus to cash crops, because subsistence agriculture was not providing income," Kashula says. "We started a pilot project in Kigoma with the support from the Mikocheni Agricultural Research Institute, which was supervising us on the production of hand-pollination hybrid palm oil seeds, and establishing demonstration sites." In addition, Tacare staff began collaborating with coffee growers in the highlands of Kigoma, working with them to access opportunities for export and marketing, and supporting coffee processing units to improve the quality of the product.

"This was a busy era for Tacare," Kashula recalls, "and the number of villages we were working with went from 12 to 24, and later, during the second Tacare phase, up to 33. We saw that our concentration was very much along the lakeshore villages, but there is an ecological connection between these and the inland villages between the ridgelines of the surrounding hills, because water sources start from the ridges in the eastern terrestrial areas." For this reason, in 1997, staff began working to increase the number of villages on the eastern side ("road side"). Kashula explains that the Kalinzi line now divides the area into four sections: Road North, Road South, Lake North, and Lake South.

"But back then during the first Tacare phase, we only had Lake North and Lake South, with the other two being added later. This was collectively referred to as the greater Gombe ecosystem. Once we were working on the scale of the greater ecosystem, we needed to consider activities at

Hafsa Ramadhan is a village nursery attendant in Mkongoro Village, where she grows passion fruit, mandarin oranges, mangos, beans, pines, and Maesopsis, known locally in Gombe as *Mshehe*. *The Jane Goodall Institute, Shawn Sweeney.*

the broader landscape level, rather than going to individual villages. In the first phase, we were concentrating on tree planting, but we had pilot villages doing individual forest regenerations that were driven by individual people, not a village-based regeneration. Working across much broader areas with the landscape level approach, we extended down into Mpanda, or Masito Ugalla. As of today, 104 villages fall under Tacare's reach, so you can imagine, it's a big scope for the program."

Even as the scale of the project increased, the fundamentals of the Tacare approach—the human element—remained unchanged. "We would still engage, listen, understand, and facilitate. Even now, I would still go and learn from the community first, answering questions like 'What do they have in terms of resources? How do they feel about these? What are their challenges?' and so on. Even when operating across broader levels like regional landscape, this personal engagement should still be a

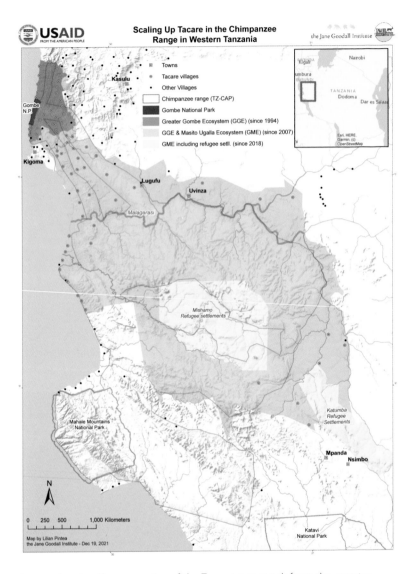

A map showing the expansion of the Tacare approach from the greater Gombe ecosystem to the larger Gombe Masito Ugalla and the greater Mahale ecosystem in western Tanzania. *The Jane Goodall Institute, Lilian Pintea.*

first step because I want to learn what their specific challenges are, because they may have some answers or some resolutions. If they do have any resolutions, we then need to understand their own capacity and from there, we can start planning together by dividing their role and our role, keeping them in the mindset of having responsibilities and driving their own progress."

Of course, Kashula says, Tacare staff still won't move fast with the assumption that the community will accept whatever they suggest. It helps to have examples of what can happen when land and resources are not managed sustainably, because the local people may not be aware of possible consequences. "One example we've used before was regarding watershed damage from tree removal, explaining that if we destroy the forest, water catchments will be affected. In one particular village, I remember, the people couldn't link these two well enough to understand, even when they had a water crisis. So they sent Tacare an application for support." As a result, JGI partnered with UNICEF to fund an infrastructure project to pipe water from the source.

JGI and UNICEF jointly prepared a proposal and, within six months, the plan was approved and funded. When Tacare first received the request, they went to the village and tested the water flow, measured in meters per second. Although the water flow was suitable for piping at that time, the local people continued with tree removal while waiting for the funding to be approved. By the time the purchased equipment reached the village, the water flow had decreased so much that the original plan could not be executed.

Kashula continues with this cautionary tale, which fortunately had a happy ending: "We formed a technical team with the government and another NGO. They prepared a report that advised us to apply this water source plan to another village. That was a learning curve for the people—how they had destroyed their chance at sustainable clean water even after we had warned against their damaging use of resources. Once the villagers knew we had to withdraw the funding support, they pleaded shame and promised to take care of the watershed if we just altered the location so they could still benefit. We allowed them another chance, and luckily they have been taking care of it since. This is one story I used as a good example for vegetation and watershed connectedness." In fact, Kashula says, he recently visited this village in Kigoma's rural district, and "it looks amazing. The water flow has increased by more than 15 times the previous measurement."

In another example, Kashula describes one of the earlier, more unresponsive villages, which was mistrustful of Tacare, as were several other villages at that time. JGI had started a nursery project at Kigalye, on the Lake Tanganyika shoreline, but after a year the community still didn't buy into the idea, so JGI continued to other villages and left it behind. The villagers refused to engage with Tacare staff to discuss erosion problems, believing that JGI would use their village land for chimp habitat and force their people out. "Because community participation in Tacare is not forced or pressured, we had to leave," he explains. "Five years later, they suffered a very destructive landslide. They came back to us to request our support in working with them again. We said, 'Not unless you get approval from the district commission, and without your participation, we cannot waste our time again in collaborating with you.'

"By then they were seeing JGI take resources to other villages, passing their village with materials for constructing schools, water schemes, etc. You know what happened? They went back to the district authorities, they got their approval letter, and they brought it here. We went back with the district; we held a public meeting during which they said, 'Now we are going to collaborate well.' Now, this village is a great model, different from other villages. When we have visitors, people from different places, we take them there, to show as an example. It's the most outstanding village in the whole program area."

While Kashula is justly proud of the program's successes—and candid about its setbacks—he is passionate about its basic principles that have, over time, evolved into a whole new approach. "Within the institute, the

Amena Hassan, a farmer and homemaker in Kasuku Village, Kigoma region, saves as much as eight hours in a day fetching water since Tacare installed a clean water system. *The Jane Goodall Institute, Jackie Conciatore.*

participatory way of implementing the programs is one reason I still have passion for this work, because as staff of JGI, we're not dictated to. We do by planning, we do by sharing, but also we do by contributing. We have monthly JGI planning meetings where everybody presents what he or she has planned and what has been implemented. From that point, you're somehow challenged to succeed, but you're also advised and supported. This kind of interactive planning and implementation is among those elements that motivated me, but the other thing is the trust."

Being trusted by the organization to try what he knows, and to share what he knows with others, has kept him motivated, Kashula says. "When we began, we were planting exotic trees to reduce pressure on natural forests. Then we realized that we also needed to reduce pressure

Kashula (*left*) facilitates a participatory mapping exercise using IKONOS satellite images and learns from local communities about their landscape and forests. *The Jane Goodall Institute, Lilian Pinea.*

on indigenous trees that were being exploited quite a lot faster than our efforts to replace them. We tried a kind of domestication, and I was charged and trusted to try my forestry skills. I came up with a device and a method that people now come to me and JGI to learn about. We managed to raise complicated indigenous tree species that have not been raised and grown by any other organization, trees like the Mninga (*Pterocarpus angolensis*)—a hardwood species that is very difficult to raise and grow.

Because of this complexity, little effort has been directed to propagate and domesticate this species. But for us, we have succeeded because the organization has trusted and supported staff like me and others to try out our skills and knowledge to innovate."

Another aspect of his work at JGI that Kashula deeply appreciates is the time he can spend learning from the local people—"because all projects we implement belong to the community, not us," he emphasizes. "Working with local people, you learn so much. For example, I'm now very familiar with a lot of medicinal plants and their application, and I've learned different practices and methods of controlling pests. I learned this from the people, from being in the villages. The knowledge being held and used by the local communities has proven feasible because it's compatible with the environment. It's also compatible with the culture of the people because it has been developed by them. It's the real knowledge that can solve the challenges the communities are facing, such as how the local medicinal trees are used to control pests. I think indigenous knowledge is a great ingredient in community conservation."

Stressing the value of local indigenous knowledge and local cultural practices—as opposed to scientific knowledge "sleeping in the cupboards"—Kashula concludes that "the Tacare approach is not like a Bible. It's something that's compatible with prevailing policy, prevailing laws. Its adaptation will depend on several factors. Tacare is always changing. We adapted an approach to integrate more than one activity, meaning that tree planting, which we began with, could not solve the problem. But with an adaptive framework, the holistic approach shifts in focus in response to community needs. We adapted by addressing people's needs first—health, income generation, education, water, land use planning. That approach has, to an extent, solved or contributed more positively toward identifying and resolving issues that ultimately lead to deforestation."

This approach, he points out, is a framework, but not necessarily one that practitioners can copy and paste. It has to be adjusted according to the needs of the people, the policies of the country, and the specific goals of sustainable development, case by case. "The needs of the people of Tanzania, or specifically Kigoma, may not be the same as the needs in Burundi or Congo or somewhere else," Kashula says. But as a framework, the Tacare approach can still be effective in a variety of places, providing it's applied in a fluid, responsive way. Like an organism, a forest, or a community, it needs to keep adapting and evolving to thrive.

Chapter 5

Cultivating a holistic approach

Emmanuel Mtiti dances with donors

Mzee Emmanuel Mtiti is the type of character who greets security guards and kitchen staff at the JGI office in Kigoma in the same courteous manner that he addresses high-ranking government officials or old friends. Always referred to by his last name, Mtiti comes across as a man with a natural and effortless dignity, supplemented by an even and measured nature. Although he is soft-spoken, he can silence a room with a single word; when Mtiti speaks, everyone stops to listen.

"I've been working for JGI since the end of 1996," he says, "so I'm getting into my 24th year with the institute." Since first being employed by JGI, he has risen through various levels of management, most recently as the program director for several phases of the conservation program across the Kigoma and Katavi regions of western Tanzania. These days, as the director of programs and policy for JGI Tanzania, Mtiti spends most of his time in Dar es Salaam as JGI's representative in government and partner forums, working closely on the coordination and policy aspects of the institute—specifically, where its work overlaps with academia, the Tanzania National Parks (TANAPA), the Tanzanian Wildlife Research Institute (TAWIRI), and other governmental and nongovernmental

Emmanuel Mtiti. *The Jane Goodall Institute.*

institutions related to conservation and the well-being of the communities. Mtiti's wisdom and contextual understanding of JGI programs and national policy make him an important diplomat, a bridge between the worlds of NGOs, local communities, and district, regional, and national governments.

"I'm from a medical and management background," he says, "and I came to know Jane very well through her involvement with the health department where I used to work. She used to come to meet with the late Dr. Mbaruku, who was the regional medical officer for Kigoma region in western Tanzania. Right from that moment, I saw her as a dynamic person who doesn't only focus on chimpanzees but also focuses on people." Jane's ability to connect with individuals, one on one, is how she draws people into sharing her passion for humanitarian and environmental issues. From Hollywood movie stars to world political leaders to children, Jane connects with people in a way that is spontaneous and sincere. It is also one of her many talents to find people outside her world and invite them into it, just as she did with Mtiti. Today, watching Jane and Mtiti interact, it's clear there is a deep rapport between them—a friendship that can switch from serious professional discussion to jokes and banter in an instant.

Mtiti was introduced to Tacare by George Strunden, the first project manager, shortly after the project began. After some exploratory meetings with George and Jane, Mtiti took the leap and decided to join the program. "I was leaving a good and permanent job," he says, "but I saw the potential in Tacare." He credits George with being a good mentor and a visionary leader who also had the ability to recruit and connect with donors. "It was through him I understood how JGI was engaged at the ground level. My participation in Tacare started by training staff on community approaches, going with them to the villages."

Right from the beginning, Mtiti was involved in designing, implementing, and evaluating the projects as they developed. "I like being at the interface, interacting with the communities, listening to the issues being raised and trying to promote their voices. It's crucial for us to bring their priorities to the forefront of our planning, and design whatever activities can best meet these and be implemented within the community. After we learned through the participatory rural appraisals (PRAs) that the key problems for them were not deforestation, we expanded outward to focus on the issues surrounding education, water, access to capital, and health. These were the main four. From the people's perspective, environmental degradation was way down the list. That really proves what Jane was thinking—that these people live in poverty, and they have a lot of family needs, all of which are more immediate than forest loss.

"We had to think, 'What can we do now? We are a conservation project, so what should we do? The EU has only given us money to plant trees.' We came up with problem and solution diagrams to better visualize the complete set of issues. We thought, 'Can we realistically make changes to improve the livelihood of these people?' The answer was no. If they have needs, for example, like not being able to take care of their own children, it's not like we can just say, 'Do family planning.'"

Of course, JGI also could not intervene and start taking care of the children in every household. The compounding nature of the problems needed to be addressed, but under the ownership and control of the local people. "We must build their capacity to do that," Mtiti says. "That is through family planning, yes, but also by transitioning the concept, explaining what it is and what it means for their future—that's all we can do. They are the ones who decide if they will or won't apply it. The main challenge then was that donors were more focused, and they work in silos, so who will be willing to fund an integrated and holistic approach?"

Projects as large as community development cannot endure without a variety of stakeholders. Professionals of various disciplines, local and national governments, international lobbyists, private donors, corporations—the list is long. Any of the stakeholders that truly have the best interests of developing countries in mind act, in a sense, as pollinators to bring about growth and change.

Consider the significance of pollinators in the natural world. For instance, in sub-Saharan Africa, the Guinea or broad-leaved ground orchid relies, like all blooming and flowering plants, on its own coalition of stakeholders. Bees, other small nectivorous insects and birds, small mammals, and even larger animals that unintentionally carry pollen between plants on their fur all contribute to the cross-pollination that ensures the orchid continues to pass on genetic material and propagate. While the pollinators receive food, the plant receives genetic couriers. If for some reason nature were to systematically remove these pollinators one by one over time, the orchid population would suffer. Similarly, community-led conservation initiatives would suffer if they had fewer stakeholders.

From the outset, Tacare and the community itself have relied on two crucial stakeholder groups to make up the coalition of pollinators for rural Africa: government and donors. A key function of the project was to educate donors to fund their areas of interest while at the same time contributing to an integrated approach.

Mtiti points out that the government of Tanzania has been vital to Tacare's ongoing progress. "We work closely with them," he says, "and over the years, we've both learned from each other through a shared passion. The regional, district, and national levels of government have all provided us tremendous support. Of course, USAID [a development agency of the US government] also supported the development of our National Chimpanzee Action Plan, which the current USAID's Landscape Conservation

Jane (*left*) and Mtiti bantering with each other during a break between meetings held at Kigoma, Tanzania. *The Jane Goodall Institute, Adam Bean.*

in Western Tanzania program is implementing, but let's not forget that it's through the sanction of the Tanzanian government that such programs can be implemented.

"The relationship between the Tacare approach and the government is one of mutual benefit. The government needs the influence of the programs to be able to sustain the lives of its people, to be able to sensitize its people about natural resource use, as well as improve their standard of living. Local people need the government as they're implementing the projects because unless the policy of the country gives them scope for application, the strategy and governance is lacking. We respect Tanzania's national governance because they're invested in building policy that empowers us to act, so for now, as they stand, they are beneficial. We take the best part of these policies to define our work—for example, the village land use plans (VLUPs)."

As Mtiti explains, the VLUP approach has proved effective in community conservation activities. Being involved in land use planning fosters a community's commitment to care for land by offering insight into competing land uses compounded by rapidly growing population. Working in tandem with the VLUPs are practices that improve people's lives, such as microcredit programs that offer people an alternative or supplementary income. Also, Mtiti says, "We now have the support of specific people like the village forest monitors, who are volunteers trained by JGI but selected and enlisted by their village governments. Our forest monitors have spent years collecting and discussing their observations with their fellow villagers and local leaders, using smartphones, tablets, and mobile apps such as Open Data Kit (ODK) and ArcGIS® Survey123. Other village champions are the community-based distributing agents for the family planning methods, the village health workers, and the village nursery attendants, all of whom took this approach to different levels."

But, he cautions, a policy such a VLUP is only effective if practitioners "take it from their shelves, dust it off, and read what it really takes to implement long-term solutions. Of course, we have come up with some modifications as users of the policy, which doesn't change the intent; nor does it affect our relationship with the government. For example, one version of the policy requires that when you start the land use planning, it must be discussed and endorsed by the villagers at a council meeting. But we realized a village has 12,000 people, so how can you gather them all for the discussion? You might get 600, or if you're lucky 1,000 come. What is 1,000 in relation to 12,000 in such a sensitive matter?

"We took it a step further, so instead of having one village council meeting as is explained by the law, we aim for hamlet meetings. This way we can target dispersed groups across various areas, which means more people participate. We get many people to participate just to make the process more democratic, so everyone is aware of what is happening. That has been one of our successes. We are trying to reach as many people as possible and we are able to report back with updates and brief overviews, so the government is kept aware and knows the issues."

JGI is now trying to replicate this approach in other countries, and Mtiti says he is proud to see how well it is performing in Uganda, in

Congo DRC, and in Congo Brazzaville. At first, he believed that Tacare would need to send people from western Tanzania to train community workers in neighboring countries, but, he says, they learned the theory behind Tacare's approach from meetings and interactions. Of course, while others may learn the theory from JGI's work in Tanzania, the practice will always be different in different contexts.

Reflecting on Tacare's growth, Mtiti returns to the crucial question of donors—recruiting donors and competing with other organizations for funding. "We have been, to some extent, trying to educate our donors not to think of each initiative as an isolated one, not to divide them into silos, but rather to work with a holistic mindset, explaining how all facets are connected under one approach. But often we still need to simplify things by keeping specific parameters of fund allocation. None of the donors have really limited us from having many partners. We only specify for their particular project, but not for the program as a whole, so we've kept many of our partners and can work with them within the context of each donor's perspective.

"Although not always the case, we do have donors now who are supporting our holistic approach and we're thankful for that—especially the EU for initiating the project funding and USAID, who have been with us since 2005, even while we were trying to introduce the holistic approach and real community-led conservation. US donors have also supported us quite a lot in that move. Along the way, we've had other donors that only wanted their funds directed toward a single subject, especially in the beginning. You can imagine trying to run a holistic project with something like four or five donors, each supporting one thing. It was very hard, because each had different reporting requirements, different systems for evaluating, different expectations, etc. We had a multitude of reports to write, some monthly, some quarterly. It was taking so much of our time and energy, forcing us to become office people rather than being the community people, the field people that we needed to be.

"If someone comes with money and only wants to support one aspect of our work, I don't think that will be our favorite selection. Of course, we will try to engage to get the donor to understand what we are doing, but if it doesn't work, then it's better we lose the money than lose the community's aspirations. There was one period, after just three years in, we had a management vacancy and at the same time, our funding renewal from the EU was rejected. We had to scale down our activities, maintaining the lowest number of staff and reducing the number of village visits and consultations just to keep a low presence. It was even difficult to operate our vehicles and boats."

Jane herself gives Mtiti an enormous amount of credit for keeping the project going during that period: "We had to write a whole different proposal and honestly, I don't know how Mtiti kept going during that year, the fourth year, but Mtiti and the team somehow did it. These are the kinds of things that people don't understand today. The Kigoma team kept it up themselves with no money coming in from the EU. That's just a huge, huge contribution to the early success of Tacare."

Providing more context to the donor defection, Mtiti explains, "Yes, we started the tree planting. But remember you're doing tree planting with poor people. I used to have one photo of a woman with all her clothes torn with her baby on her back. The mothers wrap their babies, and head out planting trees. However, planting trees does not remove their hunger or their illness. It only improves their future. There was some acceptance from people, but there was also this burden of livelihood. We commissioned rapid approval to see what the root causes of their hardships were. That was our very first survey into community issues.

"This caused problems at our end, with our donors. At this point we were now embracing the holistic approach. But none of the donors could accept that. The scientific world said no. Beyond that, our partners in the developed world said, 'You have to work inside a single function. If you are doing conservation, just do conservation; if you are doing health, just do health,' etc. That's the response we received from many donors we approached. We had to think carefully about how to navigate this."

Mtiti goes on to describe how these issues are interconnected in the communities. "If you go to the village and you're addressing someone today on, let's say, agriculture, health, or fisheries, it's all through me or another JGI team member. All these pillars of the project are being handled by a singular person from our side, but also a single person in the village will be doing multiple activities under these pillars. They're all integrated, connected in the daily realities of the people and in my work as a facilitator. The local people are smarter than we are, to the extent they take everything and mix it together. The same household is doing microcredit and fishing and is involved in tree planting. The same household, the same family, they are integrated at their household level, at their family level. Even when they divide roles, at the beginning of the day they say, 'You go do this, I go do that,' and at the end of the day, they come together—that's the integration.

"For JGI, if you are still a donor that thinks unilaterally, that will make it difficult for us to work with you. We love donors who really understand how connected all these development aspects are. Our donors need to understand what we are learning by being on the ground. Any donor who wants to develop people in one way or the other must listen to the people. We are the mouth that carries the words from the people to our donors. We have been working with them and we know what they want. It's not really that we invented this. It is the people who advised us this is how they want to do things."

It's understandable that donors would require specifics on how their funds will be allocated. After all, along with the generous act of giving comes the need to understand exactly what that gift is being used for. In addition, many donors are passionate about helping a single cause. One person may be driven to help children have access to education, while another might want to help eradicate preventable diseases like polio by supplying funds for rural vaccination initiatives. On a larger scale, organizations such as USAID understandably have their future donations written into in-house policies so that they can maintain detailed records of where funds are being allocated and ensure that targets of support are being met for a variety of needs such as health and sanitation, forestry, water resource protection, and so on.

Keeping separate silos of funding essentially comes down to showing accountability, results, and a system of metrics to manage data and reporting. Although they make sense, these are abstract requirements, developed to satisfy human constructs. They do not reflect the real world. The real world cannot be compartmentalized, nor does it follow policy or procedures. Donor money for education might provide a great service by building a school in rural Africa, but when the classrooms sit empty because the children are needed to help their parents collect firewood and farm their land for that evening's meal, then all of a sudden the donation has zero benefit. Similarly, one aspect of community development may, by its very nature, be slower to progress than others.

To illustrate: Let's say two separate donors write a check for their respective projects of interest. Donor #1 sends $10,000 to build a remote health clinic that's fully stocked with medicine, and donor #2 sends $10,000 for improving clean water access to end the recurrence of cholera outbreaks in the village. Building a health clinic is a relatively quick project to achieve, so donor #1 receives a report that shows their money has produced a wonderful, fully stocked clinic, on time and in fact $3,000 under budget. Donor #2 receives word that local bore drilling failed to find any groundwater for building a well, and that recovering the adjacent stream is the only option, which will require several years of tree planting and riverine exclusion from agriculture before the surface water begins to flow. Their $10,000 has been spent.

Having the appropriate treatment in the village health center will not stop the recurring outbreaks of cholera resulting from the unfinished (and broke) clean water project. Can the NGO reallocate the remaining $3,000 from the health clinic and buy tree saplings to help alleviate the water crisis? No, they cannot. That money was donated for a health clinic, not a clean water project. The NGO is answerable to donor #1 and cannot simply spend the balance elsewhere. As this example shows, human beings like to compartmentalize things like funding, but in the real world, all things are interlinked.

"We've had to turn everyone's mindset around in order to progress from TACARE the project to Tacare as an approach," Mtiti explains. "Although it's evolved, our general approach has remained the same, but we've had to change the terminology, mainly for donor purposes. I remember once, I went to negotiate with one donor in Tanzania and there was this comment, 'Oh, you're under Tropical Rain Forest, you may need to change it to Forests in Developing Countries, because they can't finance you from the same fund twice.' Simply, they only fund pilot projects. We had to respond to donor requirements like that, but we didn't change our approach—just changed the justification for the project.

"Initially, some donors didn't want to get involved when we requested support on behalf of our broader approach. For example, once we applied for agriculture funding. They were like, 'You're a conservation NGO, how can you promote agriculture?' 'No,' we said, 'we are not promoting agriculture in terms of expansion, we're refining and advancing the farming techniques for where it already exists.' Whether it's Roots & Shoots or Tacare, it's all JGI. That's what we stand for with donors, we tell them which are our core activities. It's great if they can contribute, but if they don't like it, we will find a way to do it ourselves. Compromising the approach for the sake of securing funds is counterproductive. Better that there be no community engagement than to separate the issues and work in silos, because we know that it doesn't work, not long term."

Simply, Mtiti says, "We believed in what we'd set out to do and whoever wanted to support us also needed to believe in it beforehand." Now, he points out, more donors and more organizations are operating in this broader space, adopting generic integrated approaches such as Population, Health, and Environment (PHE) and Population, Health, and Development (PHD). Among the first to develop such an approach, JGI's teams have always included professionals from many disciplines—foresters, agronomists, community development officers, land use planners, conservation biologists, and GIS and remote sensing specialists.

Along with this expertise, Mtiti emphasizes, community conservation requires "the people's knowledge to be included in project operations, especially indigenous knowledge. This includes giving them a platform to voice the direction they want to head after understanding their impacts on the environment, coupled with technology to make decision-making easier and relevant." Not only is technology a relatively recent resource in these rural areas, it has the power to act independently of the founding NGO, meaning that locals themselves can continue to use and benefit from these advances. The village forest monitors, for example, use smartphones and Survey123. GIS software, including ArcGIS, gives these villages a historical narrative via printed maps at different spatial and temporal resolutions from satellite imagery, which then forms the basis of community mapping. Using field-based technologies promotes a sense of ownership that might otherwise be absent if these shiny gadgets were used exclusively by NGO staff. If conservation efforts are to be effective, communities need to be mobilized for a common goal and ensure equitable benefits from sustainable management of natural resources.

"Back in the early days," says Mtiti, "we didn't have high-resolution satellite images, especially in the first six years. So that was not an alternative. We had to mark village boundaries and land tenure claims physically. As a collective of JGI team members and village participants, we walked into the forest for several days, from one point to the next. Using tins of paint and brushes, we marked an agreed point with paint to signify the boundary. This might be painted on a rock or tree. Luckily, handheld GPS devices were available back then, so we'd take these and collect the waypoint for these important boundaries. These readings had to be translated into a map drawn on the ground for easy interpretation. Luckily, when land developments came up, we were able to transfer our GPS points. Without the discrepancies of a hand-painted rock or tree that may no longer even be there, or worse, may have been deliberately moved to extend an individual's land claim, the GPS would help mediate disputes and, where needed, could evenly divide adjoining parcels of land without making any assumptions that might ignite conflicts over property ownership.

Spreading a high-resolution Maxar satellite image on the ground, Mtiti (*at left*) facilitates a community mapping discussion. *The Jane Goodall Institute. Lilian Pintea.*

"We can bring science to the table, integrate science so it is relevant and simple enough to either be used, or in the case of maps, be simple enough for the local people to recognize the features. We have examples where we brought printed maps of high-resolution satellite images from Maxar to the communities and put them on the ground and asked them, 'Can you interpret this map?' They look at it for a while and then start pointing at key areas: 'This is a place where we do our rituals.' Women come up and say, 'This is where my farm is,' and then they discuss a little bit and say, 'This is where we fetch our water. This is our burial site,' and so on. Such examples breathe life and history into a piece of paper. But when incorporating science and technology, you don't tell them, this is what you do, this is how you use it, etc. If you reach that stage, then you have introduced a very complicated science that they cannot interpret. Any science that you bring to the community has to be interpreted not by you but by the community members themselves. It's an interactive way of operating with the communities. It also creates a sense of ownership and management."

Mtiti's sense of pride and accomplishment is evident as he recalls the moment in 2015 when, on a high-resolution satellite image map, he could actually see the result of their combined efforts. "My favorite moment involving technology, specifically village maps, is when I saw the impact of our activities, comparing the before (2005) and after (2014) satellite images, acquired by Maxar at 50 cm resolution and processed with ArcGIS mapping technologies. If we'd sent someone to take photos, I would have thought it was fake, that it was altered with [Adobe] Photoshop. But when I saw the satellite imagery showing the impact of what people had been doing, the impact of our conservation efforts, and the changes over years—that was what really caught my attention. That was my favorite moment."

For a broad-leaved orchid, things are simple. It either has pollinators or doesn't. For an NGO, the stakeholders involved all have their own accountabilities, their own ideas for where their time or money should go. As Mtiti has explained, working with governments and donors while negotiating the social and political considerations is a laborious task, requiring constant management. Even with large organizations such as USAID or agencies such as TANAPA that may have a long-standing partnership with an NGO such as JGI, these organizations are made up of people. People change departments, find other careers, or maybe have a complete change of heart about their priorities. Like the orchid, JGI relies on its mutualistic relationships with its stakeholders. However, unlike the orchid that adapts over time to a reduction in pollinators, NGOs are often met with immediate funding shortfalls, labyrinthine bureaucracy, and frustrating technicalities. In the face of such challenges, every NGO should hope to have its own leader and ambassador on staff, just as JGI has Mtiti.

Chapter 6

Creating a common language

Lilian Pintea uses mapping technologies to develop a dialogue between communities and conservationists

"One of my earliest memories as a child was when dad and I would go camping," says Dr. Lilian Pintea. "We would spend a week or more away, sustaining ourselves on what we caught fishing in the lake at the bottom of a big valley. I was so excited about being out there, connecting to nature through the adventure of providing our own food." Lilian is the vice president of conservation science for JGI-USA and has been working with Jane and the JGI team in Tanzania since 2000. He was appointed to a full-time position in 2004. Lilian's field is conservation biology, and he specializes in applying geospatial technologies to habitat and species conservation, research, and community engagement in JGI program countries.

"I'm from a small town in rural Moldova," he explains, "where I was lucky enough to be a 'free-range' kid, spending time outdoors." As a child, Lilian was inspired by Bernhard Grzimek's book *Serengeti Shall Not Die* and other natural history books available in Russian, including some by Jane Goodall, which set him on a path to working with wildlife in Africa. "I loved snakes," he recalls, "and by the age of 15, still in high school, I had been studying snakes with researchers from the Institute of Zoology of

Lilian Pintea in Ugalla,
Tanzania. *Jeff Kerby.*

Moldova as part of a science youth program. At one time I remember having around 200 snakes in our small apartment to measure their taxonomic morphology before bringing them back into the wild. My poor mother, who was terrified of snakes, supported me despite all her fears. I was so lucky to have such supportive parents. But, as I was observing and learning about snakes, I was also becoming aware of their struggle to survive in an increasingly human-dominated landscape, seeing not only their vanishing habitats but entire ecosystems destroyed."

Over the years, what started as a concern to save animals and wild nature grew into a broader holistic understanding of the interconnectedness of people, wildlife, and their shared environment—and a quest for solutions. "We live in incredible times with amazing breakthroughs in science and technology. But for me," says Lilian, "a key question still is, How do we convert this amazing science, technology, data, and knowledge into wiser decisions and conservation impact? How do we stop destroying life on the only planet that we call home?"

When Lilian first came to Gombe in 2000, he was doing his PhD work at the University of Minnesota. By then, he already had more than eight years of experience applying GIS and satellite imagery, working with multidisciplinary research teams in Romania, at the University of Delaware, and as a World Bank environmental consultant in Washington, DC. A meeting with Dr. Anne Pusey, who studied the Gombe chimpanzees with Jane Goodall in the 1970s, brought him to the Jane Goodall Institute's Center for Primate Studies at the University of Minnesota as part of

Lilian (*center*) as a PhD student meeting (*left to right*) George Strunden, Jane Goodall, John Maclachlan, and Emmanuel Mtiti at the airport in Kigoma in 2001. *The Jane Goodall Institute.*

the conservation biology program. "I was the GIS guy," Lilian says, "the person with the mapping skills, which was cutting-edge technology in the conservation and primate research world at the time. I was using Esri technologies such as ArcInfo and ArcView for all the GIS needs, combined with Erdas Imagine for satellite image processing and classification." At the same time, Lilian says, his interdisciplinary studies in geography, history, and the social sciences led him to contemplate different ways of knowing and how these had shaped the way that conservation scientists had been defining conservation problems and solutions.

Soon, theory met practice. During one of his early trips to Gombe, in 2001, TANAPA officially invited Lilian to join a government survey team composed of TANAPA, survey engineers, and community members from the adjacent villages to help locate, georeference, and mark the boundary of Gombe National Park. At that time Gombe was still not officially demarcated and mapped on the ground.

The British colonial administration first designated Gombe Stream

Game Reserve in January 1943, citing its scenic beauty, threats from defor-estation, and the presence of chimpanzees. At the time, an estimated 200 chimpanzees were living in the dwindling forest areas and at risk of dying out or being displaced if the land was not protected. Shamefully, the colonial government used control methods such as sleeping sickness and other diseases to push people to live outside the area. In subsequent years, people gradually returned to living inside Gombe, only to be evicted twice more.

The government added a small amount of land to the southern boundary in 1955, and in 1968, by decree of Tanzania's first president, Julius Nyerere, the reserve became Gombe National Park. Its bound-ary was described in detail in government documents but not properly marked on the ground. Over time, both the change in protective status and the history of multiple evictions had led to tensions and mistrust between the government and the local communities.

"I would join village meetings as part of the Tacare team," Lilian says, "and local people would tell me, 'Gombe is white people's land, Gombe isn't ours.' This same attitude was the reason they were resistant to pre-serving trees or adding to the surrounding forest, thinking it was because the government wanted to expand the park and absorb even more of the community land." Because of this mindset, some locals were known to be actively hunting chimps found outside the park, for fear that the chimps' territory on village land would be attached to the national park. When the TACARE project started in 1994, the teams had to deal with all these years of history, mistrust, and pre-existing conflict.

"However," Lilian adds, "Tacare started changing that narrative by not talking about chimps but by listening to people and their needs. I remem-ber outside the old Tacare office we had a long wooden seat, and there were people sitting and waiting in the shade. I asked our Tacare colleagues why they were there, to which they responded, 'Oh, they're representatives from the villages. They have a problem, and they know that if they come to us, we'll listen.' They would come and sometimes stay and stay and stay.

Lilian (*left*) with local community members mapping land cover and land use outside Gombe using IKONOS high-resolution satellite imagery. *The Jane Goodall Institute. Lilian Pintea.*

Sometimes they would be there for so long, we would invite them to join us for lunch."

With JGI's success in building trust, the local people were beginning to welcome the engagement, and by the time Lilian joined the Tacare team in 2000, local communities and JGI were working together as partners. Then, Tacare was further precipitated by a horrible tragedy. In 2001, a flash flood started behind the hills of Mtanga village, raging down the Kizuka stream and straight through the village. People were killed, houses destroyed; later, stories would emerge of children trying to escape the flood by climbing trees, only to be washed away and never seen again. "It was terrible," Lilian recalls. "I was in Gombe with Jane and colleagues when we heard the news of the flash flood. In the morning, I visited Mtanga with Dr. Shadrack Kamenya, at the time director of chimpanzee research at Gombe, to assess, map the impact, and see how we might be able to help. What we found was a community in shock and a landscape

The day after a flash flood from Kizuka stream in Mtanga village. *The Jane Goodall Institute. Lilian Pintea.*

transformed overnight with deep erosions along this little stream, large trees uprooted, and boulders scattered, like a giant threw them across the valley.

"When Dr. Shadrack asked what happened, the locals explained, 'Well, we're sitting at night. Suddenly, we heard the drumming and saw the lights.' I was like, 'Drumming?' and they said, 'Yes, associated with the Kizuka Spirits. You see, the spirits live in trees. We cut all the trees and they got angry with us, and they just made this whole mess. This is what spirits do when they're upset. In the middle of the night, they started drumming and created this flood.' In fact, what they had thought were drumming and lights were huge boulders being moved down the hill in a landslide, causing sparks as the boulders struck one another."

Lilian and Shadrack took the opportunity to draw connections between spirits, erosion, and tree cover loss, which left the spirits angry and the village so exposed during the flood. But the villagers' response was disappointing. "They replied, 'No, the spirits are gone, that's what they

do, they make a mess and leave—so there's no spirits left in those hills.'"
Shadrack and Lilian left that day concerned about how Mtanga and other
communities adjacent to Gombe would cope with other disasters that
might come from loss of ecosystem services as the result of incompatible
farming, exacerbated by climate change. They knew that climate change
models for the region predicted more extreme events such as flash floods.

Lilian says, "We worked with the Minnesota Science Museum to help
us get the first IKONOS 1-meter resolution satellite imagery of Gombe.
IKONOS was a commercial Earth observation satellite and the first to
collect publicly available 1-meter resolution imagery from space. This was
years before Google Earth was around. Can you imagine having a picture
from space using this innovative technology and for the first time seeing
in detail trees, forests, footpaths, farms, houses, and other human land
uses in remote areas across the globe? That was quite a breakthrough."

Once JGI had access to such imagery, staff could not only map individual trees, farms, tree cuts, charcoal kilns, and other features but also
engage community members and talk about what they all could see on the
same satellite map from space. "I thought it would be important in the
spirit of Tacare to share these incredible images with local communities
and get their perspectives," Lilian says, reflecting on this pivotal moment.
"To look together into this mirror reflecting our own footprint from space,
to share and discuss what we saw. So, with support from the Minnesota
Science Museum, we printed some large maps of individual village lands
with no human-made boundaries or names, just the IKONOS imagery."

This first high-resolution satellite image helped Lilian and other
Tacare staff develop a common language through which to engage in a collaborative process of spatial planning, a dialogue with the immediacy of
a visual language. "The image of a lush and green Gombe surrounded by
bare, deforested land was important," Lilian explains, "because it allowed
us to then sit down with the local people and have a dialogue about what
was happening. They could recognize their houses, farms, trails, and other
places important to them, places of significance. It gave them a spatial

Left: A 1958 aerial photo of Mtanga village in the southern section of Gombe National Park. *Right*: A 2001 image of the same area. *The Jane Goodall Institute, Lilian Pintea.*

context, and it gave them a collective, transparent, and objective look at how their land uses had been changing their landscape." With their own eyes, local communities and government officials could see a striking contrast between the tree cover inside Gombe and outside the park.

In the past, Lilian explains, Tacare teams had tried to have such conversations using the lower-resolution (28.5-meter) imagery from the Landsat satellites. At that time, without the recognizable details in the map, local people couldn't connect themselves to the area because they couldn't identify features that were specific to them. "But the moment we shared the maps from the higher-resolution 1-meter IKONOS imagery, their own relations with the landscape became relevant," he says. "With no aerial photos available after 1974, the high-resolution IKONOS satellite imagery was the first time in almost 30 years we could share and compare, in high resolution, pictures of the village lands so everyone could see for themselves that by 2001 most of the woodlands were gone."

Today, using newer and even higher resolution images (up to 30 cm) from JGI partner and space technology company Maxar Technologies, Lilian can show that the Mtanga community has made great strides in restoring its watershed, with many steep woodland areas showing signs of natural regeneration. What Mtanga and other local communities around

Left: A high-resolution satellite image in 2005 of the Kizuka stream watershed. *Right*: A 2014 image of the same area. *The Jane Goodall Institute. Lilian Pintea.*

Gombe have managed to achieve in the 20 years since—in partnership with JGI, local government, and others—has changed Lilian's view on conservation. It has given him insight into how resilience is built in social and ecological systems, and it has given him hope.

Before long, Lilian was inspired to combine the practice of sharing satellite imagery with the practice of participatory mapping. "I knew from Tacare's use of PRAs that local people were encouraged to draw conceptual maps in the dirt or use rocks and sticks to sketch drawings to capture their knowledge and discuss their use of natural resources," he explains. "Hearing the importance of these conceptual 'dirt' maps, I realized we could combine the participatory approaches of PRAs and Tacare with actual satellite images and have local people draw their own names, boundaries, interpretations, and meanings on satellite maps. Then we could quickly 'travel' remotely through the imagery and record valuable local knowledge and culturally significant information, all georeferenced and ready to be integrated with other sources of data in Esri's ArcInfo/ArcView GIS. Not as a replacement for conceptual mapping but as a complementary tool. It's a very powerful approach if done correctly as part of Tacare," Lilian says, "and very effectively captures and records people's knowledge, values, and perspectives."

To illustrate, he points to a photograph of community members standing around a map printed from a satellite image. "As you can see," he continues, "the women are all standing back, or being pushed back by the men. But of course, women are the ones who do much of the work and have a different and more detailed knowledge of landscapes and natural resources than men. They farm, they collect firewood and water. So, we have focus groups which are designated by gender, age, and other social and economic factors; this way we can capture the knowledge from multiple perspectives. One woman explains, 'This is my house. This is the footpath I walk to get to my field. This is the tree where I put my baby in the shade under the tree to sleep as I'm farming. This is where I go to collect firewood, and this is how I go back home.' This is GPS-precise information of this woman's use of landscape in a day and how she values and relates to that landscape."

Satellite imagery connects Lilian's world of science and technology with the world of biodiversity and its custodians, capturing the complex relationship between them. "Suddenly, we could look at the same map and discuss forests, people, and chimpanzees," he says. "It became a common language between science and local people. That was really important because it built organic relationships with these villages, as they were involved in the planning, the science. Equally importantly, their daily lives and needs could be formalized and made relevant in the eyes of outsiders."

Using this integrated information from local communities and scientific research, JGI worked with an interdisciplinary team of consultants to study the conservation outcomes of the first 10 years of Tacare. Lilian was part of that team, funded by USAID, and he supported it with satellite imagery classification using Erdas Imagine and Esri's ArcInfo/ArcView GIS for spatial analysis and mapping. The 10-year report was an important milestone for JGI and Tacare, helping it adapt and evolve into the process of active listening, planning, and action that now we call the Tacare approach. The assessment confirmed that after 10 years of operation, Tacare had been successful in opening the doors to communities and

Mama Sania (*back right*) facilitates a 2003 women's focal group for community mapping using IKONOS satellite imagery. Lilian (*top left*) takes notes. *The Jane Goodall Institute, Lilian Pintea.*

had contributed to increased awareness, positive attitudes, and practices that could benefit long-term conservation. However, the report concluded, Tacare needed to be more strategic and spatially focused on prioritizing the most direct threats to forests and chimpanzees.

As a result, JGI partnered with The Nature Conservancy (TNC) in 2005 to adopt, for western Tanzania, a systematic conservation action plan (CAP) approach widely used by TNC in North America and other regions. The goal was to explore how the Tacare approach could better connect community livelihood efforts with conservation objectives and outcomes. Building on Tacare's 10 years of success and partnership with the local communities, the CAP process enabled scientists, local government officials, and other stakeholders to engage and collaborate, informed by expert knowledge, science, and local community data. Through this planning work, JGI gained valuable insights. For example, one of the first initiatives requested by the communities, and supported by Tacare, was planting trees to address the rapid increase in deforestation on the village lands outside Gombe National Park. By 2004, Tacare tree-planting initiatives had spread to 14 villages outside Gombe, but, despite these efforts, satellite images showed the deforestation rate in areas important to chimpanzees had almost doubled, from 87.5 hectares/year from 1972 through 1991 to 171 hectares/year from 1991 through 2003.

Tree planting was intended to provide local people with an alternative source of timber for boat construction and household consumption. Yet the CAP process revealed that the main threat to forests outside Gombe was not logging for timber but farming. Most of the forests and woodlands on the steep slopes were converted to subsistence agriculture, with more fertile valleys devoted to cash crops. While tree planting and nurseries were still important to support people's livelihoods and need for timber, additional strategies were needed to address the main drivers of deforestation and the main threats to forests and habitats.

Equipped with this new knowledge, Tacare and the JGI Tanzania team used the greater Gombe ecosystem CAP to adapt and implement new strategies focused on farming and land use. The land use plans were developed collectively by the local communities and government, guided by Tanzanian land policies. JGI facilitated the process by providing resources, with support from USAID and with geographic focus and guidance from the CAP and geospatial technologies in partnerships with Esri and DigitalGlobe (currently Maxar Technologies).

By 2009, 13 villages had completed their participatory village land-use plans and voluntarily assigned 9,690 hectares, or 26 percent, of their village lands as village forest reserves. These reserves were interconnected across village boundaries to minimize fragmentation, covering 68 percent of the priority conservation area delineated by the CAP. The next step was to help local communities build their capacity to implement and enforce those land use plans and forest reserves. Drawing on Tacare's experience, JGI Tanzania and the local communities came up with the idea of village forest monitors—community volunteers selected by their village

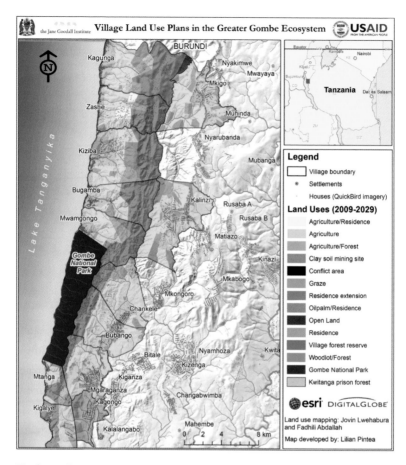

The first village land use map of communities outside Gombe National Park developed in 2009, showing the location of the interconnected village forest reserves set up by local communities and connected to Gombe National Park, guided spatially by the greater Gombe ecosystem CAP.
The Jane Goodall Institute, Lilian Pintea.

government. Once the volunteers had been selected, JGI offered support in the form of training and access to knowledge, technologies, and tools.

At first, the monitors' data collection was via written notes and paper protocols, which imposed limitations on managing and using the community data. When handheld GPS devices became available, project staff introduced Garmin GPS devices, and in the first year, the forest monitors collected more than 30,000 GPS waypoints. However, these simple GPS devices were unable to capture the variety of attributes needed to inform decision-making; the monitors needed an alternative solution that could gather more information while also providing a simple way for data to be organized and accessed.

Over a four-year period, Lilian and the Tanzanian team reviewed CyberTracker software running on Palm Pilot devices and the even more sophisticated Trimble GPS units. But none of those approaches met JGI's needs to provide technology solutions that were both cost effective and user friendly for non-GIS users, such as village forest monitors in the field. Finally, in September 2008, the first Android smartphone, T-Mobile G1, was announced by Google. Google Earth Outreach, the Amazon Conservation Team, and Chief Almir Surui then invited Lilian to Brazil to see how the Surui tribe of the Amazon were reporting their field observations using the new Android smartphone and a free and open-source mobile app called ODK. Impressed with these emerging technologies, Lilian, with support from Google Earth Outreach, introduced Android devices and ODK into the village forest monitoring program in 2009. In preparation, the JGI team sat down with the forest monitors and posed questions such as: What type of information did they think they needed? What wildlife signs were they able to identify and interested in collecting? What illegal activities in their village forest reserves was their village government interested in monitoring?

"After listening to their information needs, we would develop a protocol and an ODK form, upload the form to the Android smartphones and tablets, and train the village forest monitors to use ODK to collect the information deemed most relevant to their own communities," Lilian explains. "We didn't come with an agenda saying, 'JGI wants to collect this data.' Instead, the data collection needs are defined by the communities themselves. And those needs have evolved as village forest monitors have been learning more about the data and what is actionable for

Village forest monitors test the first Android G1 mobile phones and ODK mobile app in 2009 in Kitwe Forest, Kigoma, Tanzania. *The Jane Goodall Institute, Lilian Pintea.*

their communities. We listened to that feedback and adjusted the ODK forms as needed every time we did the training. Of course, as one of the first such programs in Africa, we learned many lessons along the way. For example, how to provide access to power—these villages did not have electricity. How to provide access to the internet, and how to avoid destroying mobile devices in extreme humidity, heat, and dust. All that had to be figured out by trial and error."

Five main categories of data are collected today: monitoring or patrolling efforts, evidence of wildlife presence, more detailed information on chimp presence, evidence of illegal activities, and an open form to record anything that village forest monitors think is important to note. The latest addition to the list, in 2019, was reporting evidence of wild bees, wild beehives, and mushrooms.

As Lilian explains, "People said, 'Oh, we really want to map mushrooms.' I was wondering, 'Why mushrooms?' And they said, 'Well, as some of our forests are restoring, the mushrooms are coming back, and we appreciate it so much.' I also was wondering why they wanted to map wild bees and beehives, and they said, 'Well, we want to map wild beehives because a forest that has native bees is a healthy forest. It's an indicator of habitat health,' which was great to hear and learn about."

2019 was also the year that JGI transitioned from ODK to ArcGIS Survey123. "As a mobile data collection app, ODK served us well for 10 years," Lilian says. "But, as we've been expanding our outreach in Tanzania from 12 to 104 villages and scaling up community monitoring efforts across Uganda, DRC, Congo, and Senegal, we needed a data collection app integrated with an entire GIS system to help JGI manage, analyze, visualize, and share this data. Survey123, which utilizes the same open standard that underlies ODK, enabled JGI to leverage the full power of ArcGIS software. It means that the community data automatically connects to the JGI geodatabase in the cloud and generates dynamic dashboards, updatable in real time, and dynamic web maps that can be organized as part of specific decision support systems. Integrated full GIS platforms like ArcGIS are really useful for organizations like JGI," he notes, "because we can focus on using the data to improve conservation decisions and impact, not spending our time trying to get different systems to talk to each other."

Lilian points out that the way Tacare is leveraging mobile technologies with local communities is not typical citizen science. Citizen science projects are often crowdsourcing efforts focused on generating better data for research. But when local people voluntarily commit their precious time and participate in collecting data on forest health, chimpanzee sightings, or native beehive locations, their participation induces community ownership and trust in the data. It also gives these communities a voice, which they then use to communicate with local government.

The challenge is still to understand how this information trickles down into actionable solutions. At the community level in western

Tanzania, decisions are often made through a dialogue involving elders, village government officials, representatives of community-based organizations, village council members, and other community leaders. In this context, a georeferenced picture shared on a smartphone can be useful to inform the community dialogue.

"For example," Lilian says, "one day a forest monitor from Kigalye village took a picture and georeferenced the location of a bullet cartridge that he found inside his village forest reserve. I saw the photo and was wondering why he decided to take a picture and not just record the location of this evidence of hunting, an illegal activity in the village forest reserves according to the village bylaws. At the time, because of limited internet connectivity and access, we encouraged forest monitors to only take pictures of important events that would benefit from a photo, to minimize the size of the data to be uploaded. But he said, 'You see, this is the first time I found a bullet cartridge on our village land. It seems because our village forests are coming back, animals are coming back, and with animals, poachers are coming back as well. I took this picture to show my village government that we have a new problem in our village forest reserves and that we need to do something about it.' So, when properly used within a decision-making context, one picture can be worth thousands of words...or thousands of journal articles."

For JGI and other conservation organizations, the challenge is to bridge the gap between scientific data and actionable, appropriate solutions that lead to wiser decisions. The situation is urgent: chimpanzees, like all other great apes, are facing extinction. An academic paper publishing the latest data and results is of little value if that information is not used, for example, by communities to develop a smarter land use plan that works better for people, animals, and the environment. Having a stockpile of scientific knowledge locked in peer-reviewed publications and not incorporated into conservation practices is of little use to communities or animals at the edge of extinction. Such knowledge, in Kashula's words, is "just sleeping somewhere in the cupboards."

To Lilian, the obstacles in his path are clear: "We have a 'last mile' problem in conservation. The future of biodiversity is in the hands of local communities. While we've been adopting amazing new technologies, platforms, and tools, and generating impressive amounts of data and knowledge, what's much harder to achieve—and has a much higher cost—is building capacity and providing resources for local communities around the globe to use this data and these tools. Another challenge, we've discovered, is that data, information, or knowledge are minor factors in influencing human decision-making. We know that larger amounts or better data and information products don't inevitably lead to increased use by decision-makers and ultimately to conservation outcomes. It's important to consider how these technologies, knowledge, and tools are perceived and filtered through existing local beliefs, traditions, values, experiences, and concerns, so that they're trusted and used by the local decision-makers."

Lilian has spent decades trying to make this happen for JGI. His use of GIS with local communities in some of the most remote regions in Africa has shown him that the Tacare approach creates spaces for dialogue in which data and knowledge are not simply transferred to local stakeholders. Rather, these encounters involve a process of negotiating new meanings, where researchers, community members, government officials, and other partners engage, collaborate, and understand together how data, information, and knowledge are produced—by whom and for whom. This has enabled JGI and partners to develop geospatial applications and solutions with—and not for—the local communities, as part of locally driven development and conservation efforts.

Tacare recognizes that conservation is a local social and development process that engages science, not a scientific process that engages society. Through a simple visual language such as high-resolution satellite imagery, complex data and meanings can be shared and instantly understood in the form of a map. Likewise, the local people's knowledge, land-use practices, culture, and spiritual relationships with the geography around them provide Lilian, JGI staff, and global audiences with insights into the realities, values, and meanings of life in these regions. Weaving science and technology into the context of community development and its countless variables is indeed a hard union to officiate. But, asks Lilian, "What's the use of data, knowledge, or innovative technology if these aren't making their way down to where the impact of decisions is felt the most?" At first glance, village life and geospatial technologies may seem worlds apart. But, when combined through the Tacare approach of participation and open dialogue, they begin to create a common language for working together toward resilience and change.

Chapter 7

Local ambassadors: Learning from and speaking for the chimps

Gabo Paulo Zilikana, Eslom Mpongo, Hamisi Mkono, Fatuma Kifumu, and Yahaya Almas reflect on decades of chimp observation at Gombe

As part of the longest continuous African great ape research program in history, the chimpanzees of Gombe have contributed significantly to the fields of ecology, evolution, and behavioral science. Jane's earliest discoveries are still the subject of follow-up research today. Through her observations and publications, this population has brought previously unknown chimp behaviors and characteristics to the world stage. We now know that chimps modify sticks and stems to make tools for termite fishing, they form coordinated hunting parties and prey upon other primates and vertebrates, they can be cannibals, they form deep mother-infant and sibling bonds, and they show empathy and compassion toward other individuals. They've also shown that mutiny, political alliance, and social organization strategies are all part of their daily lives.

The legacy of these discoveries has led to detailed records of some 350 individual chimps, generated more than 300 research-based publications, lured more than 250 individual researchers to Gombe, and spawned

Clockwise from top left: Gabo Paulo Zilikana, Eslom Mpongo, Hamisi Mkono, Fatuma Kifumu, and Yahaya Almas. *The Jane Goodall Institute, Adam Bean and Michael Wilson.*

roughly 50 PhD and master's qualifications. For the most part, however, these have been the findings published by foreign scientists continuing Jane's legacy in the search for answers. And these discoveries were made possible by the profound and unpublished understanding developed by the many local people who have been employed as field research assistants for decades. Although the study of chimpanzees at Gombe predates the launch of TACARE by many years, both projects are founded on valuing the skills, knowledge, and determination of the local people. Even now, at Gombe, field researchers from nearby villages carry out the task of following and observing the chimps from dawn till dusk every single day, alongside Tanzanian scientists working closely with their partners in academia to continue Jane's vision.

Trackers Iddi (*left*) and Samson (*right*) study a chimpanzee named Titan in Gombe National Park, Tanzania. Local researchers follow chimps daily, recording social, demographic, and behavioral data from the moment the chimps wake up to the moment they build their next evening's nest and go to sleep. *Nick Riley, 2010.*

Gabo Paulo Zilikana is one such field researcher. Towering over the average person by head and shoulders, Gabo could easily be mistaken for a professional basketball player on vacation. A man of few words, he has spent an incredible three decades in the forest with the chimps. "Even though it is difficult," he says politely, "I am doing this job because the founder, Dr. Jane, started the research because she had the ambition of building a conservancy, to protect the chimps, the forest, and everything in it." Still active in the field, he has witnessed significant changes over the years.

"Today," Gabo continues, "I can say that the chimpanzees' situation is good. But when I started working in 1971, there were three groups of chimpanzees. However, the big group that was in the south has been decimated; it is no longer there. Now we're only left with two big groups.

Looking at how the chimps live now, it's different from how they lived up until the 1980s—even their leadership and reign is different. Those of that era used to exercise great power to rule and fight other males, but now, life for the current high-ranking chimps is very easy. Even so, some social behaviors are very different or more common compared to what I saw in my earlier years. For example, the alpha male ruler deciding to kill an infant chimpanzee. We are always learning something new because they are constantly changing their behavior."

Gabo notes that it's because of the chimps that the Gombe area was granted protected status in 1968. These days, he says, the people around the park understand its value much more than in the past. Previously, they were entering the park for what he calls "illegal activities," but as their awareness of the forest's importance has increased, their attitudes and practices have shifted too. He attributes this cultural shift to JGI's collaboration with TANAPA on the one hand and with the villages around the national park on the other. "Today," he says, "people even strive to provide information or inform JGI when they discover that an illegal activity is being committed."

Field researchers like Gabo also provide security, he points out, "because we spend our entire day in the wild. Poachers know we're moving through the forest, so our presence alone helps provide protection." Even so, they do still see problems in the field, some of which are beyond their control. These incidents are reported to TANAPA rangers, who respond quickly. "For example, if we discover traps or illegal entries, we provide that information."

All these levels of protection have shown to benefit the ecosystem and the overall health of Gombe. "Up until around five years ago," Gabo notes, "some of the rivers outside the park on village lands were either dry or showing signs of drying up, because the hills had been left bare. Now, the hills have trees and the rivers have begun to look like they are coming back to life, though still not as they were in the past. My hope for the future is after I leave here and retire, I want to continue to provide

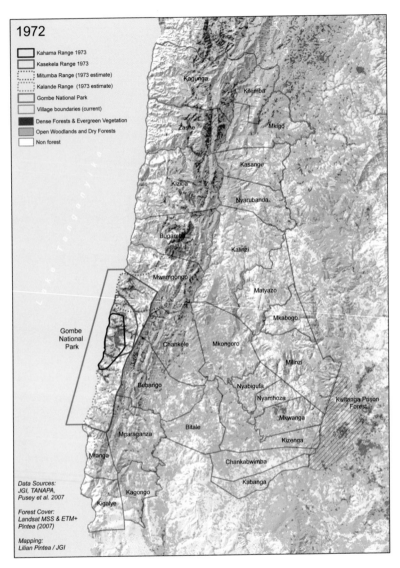

Forest and woodland cover along the shore of Lake Tanganyika in 1972.
Gombe National Park is shown in the lower left, along with the four
chimpanzee community ranges in the park at the time. *The Jane Goodall
Institute, Lilian Pintea.*

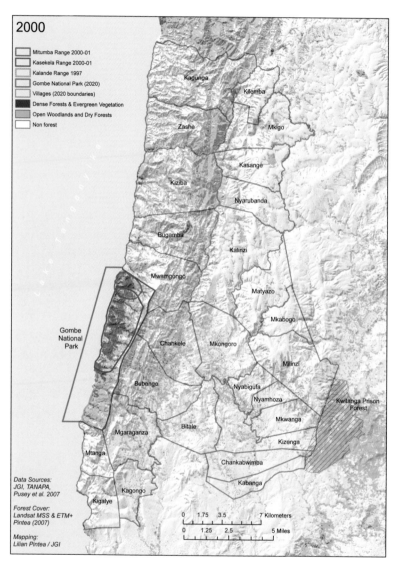

2000

- Mitumba Range 2000-01
- Kasekela Range 2000-01
- Kalande Range 1997
- Gombe National Park (2020)
- Villages (2020 boundaries)
- Dense Forests & Evergreen Vegetation
- Open Woodlands and Dry Forests
- Non forest

Kagunga

Kilemba

Zashe

Mkigo

Kasange

Kiziba

Nyarubanda

Bugamba

Kalinzi

Mwamgongo

Matyazo

Mkabogo

Gombe National Park

Chankele

Mkongoro

Milinzi

Bubango

Nyabigufa

Nyamhoza

Kwitanga Prison Forest

Mkwanga

Bitale

Mgaraganza

Kizenga

Mtanga

Chankabwimba

Kabanga

Kagongo

Kigalye

Data Sources:
JGI, TANAPA,
Pusey et al. 2007

Forest Cover:
Landsat MSS & ETM+
Pintea (2007)

Mapping:
Lilian Pintea / JGI

Lake Tanganyika

0   1.75   3.5   7 Kilometers

0   1.25   2.5   5 Miles

The same area as shown on page 90, in 2000. Forest cover has decreased dramatically outside Gombe National Park but has increased within the park. Gombe chimpanzee community ranges are now limited to areas within the park. *The Jane Goodall Institute, Lilian Pintea.*

education, especially to young people who are not yet aware of the importance of conservation—what it means to conserve the forests."

Gabo is still an active field researcher in Gombe, yet many others have preceded him, most of whom are now retired. Eslom Mpongo, now in his 80s, is one of those retirees, having left his post when advancing age made it more difficult to get around the extremely mountainous terrain of Gombe.

Born in Bubango, Mzee Eslom moved to a village adjacent to Gombe in 1971, frequenting the park to catch fish. Shortly afterward, he was recruited to the chimp research team. "I learned how to become a field researcher after originally being a porter," he explains. "We would carry bags of equipment and food for the white researchers into the wilderness." Although Mzee Eslom's eyesight is now all but gone, he closes his eyes and inclines his head, as if to visualize his early days as a fit, able-bodied young man. "I soon became a field researcher. It seemed like a good job, and because I'd done three years of schooling as a boy, I knew how to write. I couldn't go beyond third year with my schooling as a child because my parents had no money. Still, I knew enough that I was chosen to be a teacher to teach all those who could not write."

In 1975, 40 armed men crossed Lake Tanganyika from the DRC (then Zaire) and stormed Jane's camp in Gombe. The rebels, who were trying to overthrow Zaire's government, took four foreign students as hostages and kept them for many weeks, releasing them in stages as negotiations and ransom payments ensued. The result for all those involved and for Gombe as an active research center was devastating—so much so that no foreign researchers or their students were permitted back until 1989, some 14 years later. Apart from Jane, only locals like Eslom were able to carry on with the research and daily observations of Gombe's chimps. "It remained as our work because they all left," Mzee Eslom explains. "All of the work was up to us. It remained our job. When someone tells the history of the research and where it came from and where it was, it was a very difficult task. The work was hard because at that time, we were sleeping in the wild because of the kidnapping case and everyone feared for his life. But because we were all committed, the work still continued." Without the data that they collected during those 14 years, continuity would have been irreparably lost, and vital information would never have made its way into academic publications.

"It was a long time ago," Eslom says, closing his eyes again. "Almost everything I saw the chimps do, I thought 'This will remain with me for life,' but now as I sit here, I realize much has been lost. I remember Fifi, Figan, Gigi, Passion, Flo—I spent a lot of time with these chimps. It was impossible to hate anyone, although there were the ones with different personalities." Like all the others who came after him, Eslom speaks of these chimps as though they were people. He adds, "Figan was my favorite, I named my own child Baby Figan, out of love. I loved the chimpanzees."

It is understandable that the local researchers, spending endless days watching and recording their behaviors, should become attached to the chimpanzees. Also, chimps by their very nature are engaging subjects. Peter Jenkins, of the Pandrillus Foundation in Cross River, Nigeria, put it best when he said, "If you want to learn more about yourself, and about humans in general, just sit and watch chimps." Perhaps this is why people like Hamisi Mkono, who joined Gombe for no real reason other than steady employment, wind up being captivated by the chimps, taking their commitment beyond that of a normal job.

"I was born in 1941, in a place called Buzubu, in the Mgaraganza region of western Tanzania," Hamisi says in a husky voice. "I began at Gombe in 1972 and retired in 2005. I started working in Gombe as a cleaner, doing laundry for the students who were studying there. At the time, there was a young chimpanzee brought to us by tourists that had rescued it from elsewhere. When the chimpanzee first arrived, there was a young lady who was taking care of the youngster. After she left, I continued to raise the chimpanzee together with a white lady whose name I've now forgotten, until he became bigger. We started recording what he was doing. After that, it came to a point where the wildlife services intervened, saying that animals should not be cared for in protected areas. They were going to take him to Kenya. He loved me so much and we lived with him for so long, I was told that because the chimpanzee is leaving and I was his friend, I could accompany him on the journey. But when it came to say goodbye to my parents, my mother said I couldn't go—she was sick, and I needed to stay close by. The chimpanzee was taken away, and I was told that I could go back to Gombe." Hamisi says this with a stoic expression, as if trying to conceal a sadness that persists some 47 years later.

Hamisi returned to Gombe and started his new role as a researcher, studying for about three months. "After my training, which was difficult, I realized this was a lifetime job. I came to know Dr. Jane, and I really thought of her as an expert. She was able to live there, go into the forest with the chimps and just stay there." As he expresses his admiration, Hamisi still seems almost surprised by his own words; in a patriarchal culture such as Tanzania, especially back in the 1970s, the idea of looking up to a young woman of roughly the same age is quite a stretch.

"As a representative of the community outside the park," he continues, "I think the contribution of Gombe research has brought us tremendous benefits." Hamisi says he had to counsel his relatives and coworkers to understand the conservation work as something that would help them. "Also," he says, "I was trying to teach people about these animals, because local people didn't really know or didn't care."

Hamisi, on the other hand, knew the animals intimately. "For those of us who lived with them, there weren't any naughty chimps, apart from Humphrey. Humphrey was the naughty one." The term "naughty" is used by the locals to describe belligerent chimps. "Even towards human beings, he had an attitude and would sometimes throw stones or threaten to beat us. Later, after Humphrey left, there emerged a young man called Frodo. He was also naughty. He started his naughtiness while still young just like a game, but this later became much more serious.

"Chimps like Passion, for example, that chimp was a witch," he says casually. "The way she was, her behavior, the way she used to cooperate with her children to plan and kill the children of others." Witches are part of the local belief system in many rural Tanzanian villages, hence Hamisi's matter-of-fact tone.

At this point, Mama Fatuma Kifumu, a renowned local character, joins the conversation, wanting to share her initial impressions of Jane Goodall. "I'm 70 years old, I think," she says. "Maybe I'm older. Yes, I'm between 78 and 80 years old. We moved to Gombe with my husband and honorable elder, Mzee Iddi Matata. He became the leader of all fishermen, which meant he had to be consulted on anything that was happening. We lived there for some time, then a white lady came; I have forgotten her name, Jane? She found us through her search for chimpanzees. She would wake up at six o'clock in the morning and go into the forest until six o'clock in the evening. She would even get rained on in the forest.

"In the evening, all fishermen would go home, and we would remain behind. She would ask about everything. That white lady would not rest, and we would pity her: 'Is this someone who can go and stay there the whole day and get rained on?' Sure enough, she was always out there. We knew she had a job to do, just like everyone has their own duties. She worked there tirelessly, and she got what she wanted. If you seek, you will find."

Like Mama Fatuma Kifumu's husband, Mzee Yahaya Almas was originally a fisherman, based in the Kigoma rural district, relying on his daily haul from Lake Tanganyika to subsist. He learned of employment opportunities at Gombe through word of mouth and began working there in 1971. "News of the research came from our elders, the late Mzee Hilali Matama and Mzee Rashidi Kikwale," Mzee Yahaya says. These two characters featured significantly through Jane's early years in Gombe. Mzee Rashidi Kikwale was the father of Jumanne Kikwale, introduced earlier in the book, and the first person to guide Jane into the forest to track chimps in 1960, while Mzee Hilali was Jane's first official research assistant, hired by the Gombe Stream Research Center in 1968.

"After Mzee Hilali joined the chimpanzee research center," Mzee

Yahaya continues, "I asked him if he would consider me for a position. He agreed and helped me apply for a job as a laborer. After working as a laborer, I began as a slasher, slashing the forest trails for those who were following the chimps. This is where I first saw them and developed a real interest. I said, 'Brother, I love chimpanzees. I don't know how, but I want to work with them.' As the senior field assistant, he agreed to support my interest and found a place for me in the team. The ones who taught me were Hamisi Mkono, and Mzee Hilali and Mzee Eslom." Today, Mzee Yahaya speaks with great passion about the Gombe chimps. Although he may have come to Gombe simply to seek employment, his work there quickly matured into a deep and genuine enthusiasm for conservation. Not only that, but he is a natural storyteller, with a wealth of observations and anecdotes from his decades among the chimps.

"It felt like I was at school," he explains, beginning his long story with an animated expression, "like trying to remember all the knowledge and information about chimps, learning chimpanzee behavior, but also learning human behavior as well. I was so fascinated by the nature of the chimpanzee, the behavioral similarities between chimps and humans. There's only a slight difference between us. I don't know how much exactly, but because we spent so much time with them, we came to know each other like family. They know, this is our friend. Then when a chimpanzee loves me, when walking with him in the wild, he is like a brother. Yes, he's an animal, he's a chimpanzee, and you're human, but he is a brother."

Mzee Yahaya says that, while he was grateful to be taught by his elders, the chimps were also his teachers. "First, I was taught to observe, so that I could see the lessons the chimps were teaching me. So human and chimp lessons, they go together. For example, I used to watch the chimps medicate themselves, by choosing certain plants in the forest. They would go and pick the leaves from a fig tree, and roll it with their tongue, chewing on it little by little. After they had swallowed it, I would follow them and later see they have defecated worms. Then I realized this was medicine. If you follow chimpanzees, you can see they know their medicine;

they know medicine very well. All you have to do is watch their actions carefully, and soon you'll know what lesson they're teaching. But there are some things you can't see. For example, there was one chimpanzee, Humphrey; he did not know me well, and I did not see his dislike. I learned the hard way, because he took a stone in each hand and threw them at me." Mzee Yahaya stands up at this point, throwing two imaginary rocks in an uncanny imitation of chimp movement and mannerisms. "Dr. Jane said they might do this, but I did not see the signs."

On the other hand, Mzee Yahaya says fondly, chimps like Freud and Prof were his brothers. Those two were fond of playing games to trick each other. "I remember one day, Prof was hiding in the bushes, making a thumping sound to make Freud curious," he recalls. "When Freud went over to investigate, Prof ambushed him to the ground and they both ran laughing. They played that game often. Prof, he was my favorite. He would walk with me to protect me from harm. He once saved me from a snake I didn't see, by running over and shouting to warn me. He was standing two feet in front of me, shouting, because he knew the danger and knew I might be killed."

Something else Mzee Yahaya observed is that chimps are "like humans in matters of governance—it's the same as us. For example, when I began working there, the chimp's leader was Figan, and little by little, young Goblin appeared. Goblin had no older brothers, he was just a little boy, but he had the sense to follow Figan, because of his status. He was trying to learn how to take over the leadership, so he began to associate with Figan. As Figan was building himself up, Goblin followed him to watch and learn." The researchers use the term "building himself up" to explain the dramatic displays of posturing, strength, and speed males use to show group members how formidable they are, should others be tempted to challenge their ranking. "Goblin sees, 'So this is the way he builds himself?'" Mzee Yahaya continues. "Goblin copied and began building himself, and when Figan calmed down, Goblin went to pay homage and respect by holding his back and rubbing it, just like hugging. That was how Goblin kept his alliance, and he stayed that way for many months. Some time later Goblin realized, 'I can overthrow him. Have I not studied him?'

"Goblin had learned everything he needed to learn to win a battle. One day, Goblin was building himself up, so Figan passes in front of him. Then Goblin passes in front of Figan in response. Suddenly they rush each other and collide, fighting, wrestling, only the two of them. Twice, then thrice, until Figan was defeated by Goblin. Afterwards, Figan went over to Goblin and acknowledged his dominance, and I was surprised to see young Goblin take control. He was small, and he had no male kin. Figan saw that he was being ruled by a small child and eventually left the group. We followed him, said goodbye, and he was no longer seen, he disappeared. Of all the researchers, no one could find Figan, or see if he later died—we didn't know. Under Goblin's reign, we could see that Figan's sister, Fifi, hated him. She was very hurt to see Goblin take control of her brother's rank; she hated it a lot. She was a great enemy to Goblin.

"Goblin continued to reign, but he still had no friends. Because of this, he was vulnerable; he had no support from any of the other group members. With no allies, Goblin became suspicious that maybe Freud would challenge him. Without waiting, Goblin violently attacked Freud, bit and tore at his limbs, causing deep injuries so that he couldn't challenge and take control. He knew his throne was just borrowed, that they would come and attempt to rob him later. One day when I was on duty taking observations at the banana feeding station, Goblin appeared and as he emerged, he found Fifi there in the camp, sitting quietly. Goblin sounded like he was begging to Fifi a little bit, trying to get her approval. But Fifi didn't like it, so she sounded an alarm, a dangerous voice. Goblin tried to beg her, but Fifi kept shouting. Goblin became distressed because he already knew what was going to happen. Soon the other males arrived. They went from one group to the next, stomping on the ground to gather support. Even Fifi stood and followed them.

"Goblin tried to beg them, but nothing he did worked. Wilkie led the charge, and they rushed to beat him. Goblin climbed a tree, but they

Male chimps in mid-fight during a political display of power, strength, and social status. *The Jane Goodall Institute, Jane Goodall.*

pushed it around until he fell out, then he ran. I followed him myself since I was on attendance. They went northeast up the trail to the mountain pursuing him—they all went to ensure he didn't escape. They caught up with Goblin, built themselves up, and attacked him, around six chimpanzees sitting on him and punching, gnashing their teeth on him until they finally let go. I followed him up on the cliffs, in the early evening where Goblin finally settled and slept alone. He had serious injuries.

"As soon as Goblin was cut down, when they hurt him, he'd already lost his leadership. Wilkie took the lead, and after 18 months, he handed the superior position over to who? Freud. Freud had returned. It came full circle back to Freud. This is some of the governance and politics I found to be very fascinating because most of the time, they are very peaceful and loving, especially the females. But they can also be ruthless if the rankings are not in balance."

Here, Mzee Yahaya pauses for breath, but soon embarks on another story about chimps' hunting behavior, an attribute he hadn't known about before he started observing them. "The way chimpanzees hunt—they consult each other beforehand. For example, if they are walking in the forest as a whole group, they begin to consult, knowing they have passed a certain animal, and then the male releases a sound as if awakening the hunt. They have a certain voice, that is, a special signal to get ready. They turn around and look at the females and their infants. After placing their infants on their backs, the females climb into the trees, and the males start the hunting march. When we researchers saw that, we knew there is danger, and you must stay on the sidelines. We would need to move around in complete silence.

"When a chimpanzee walks in the bush during a hunt, he will move a single leaf aside just so he can place his foot in complete silence. He walks very slowly, and very carefully. He does not utter a single sound. When they walk, you stay silent and just listen. Afterwards, you hear calls and know the pursuit has started. You run towards the sound of chimpanzees and find they have captured the piglets. The mother pig attempts to save her young, but she fails because she cannot follow them into the trees. The chimps eat the piglets from up there. So, chimpanzees hunt just like humans; they communicate a lot, and there are specific sounds they use. I love to observe chimpanzees and their behavior, especially when they communicate. And if you are a long-term researcher, you will definitely come to learn what they are saying, what has just happened, and what is about to happen.

"Today, I miss the chimps," Mzee Yahaya says, with obvious sadness, especially in contrast to his animated impersonations of ape battles and piglet-eating. "I was very disappointed to retire. When the old man leaves, things change. It's a difficult issue, and I even think about the chimps, what they thought when we retired. The chimpanzees knew we were their researchers. Would they think, 'Where have they gone?' To this day, and no lie, to this day, there are times when I sleep and have dreams, I see chimps and I follow them. To this day. I wake up at night and say to my wife, 'Mama Bahati, I sleep, and I see chimpanzees; I follow chimpanzees. How did this get to my head?' To this day, it still follows me."

After he left his job at Gombe, Mzee Yahaya bought some land for crops and a banana plantation. "I built my small house and that's where I stay," he says, wistfully. "I would love to be in the forest again with the chimps, but I can't do that now, I can't. My job is to enjoy the Gombe view from here and that's what I tell my teens, Magombe and Amri. That was a job for one day, every day, but the experiences I had are lifelong." Pausing to reflect, Mzee Yahaya emphasizes how grateful he is for having been employed at Gombe, and how much he believes the Tacare approach has benefited the communities in the surrounding villages. Then he shares some precious memories of Jane Goodall.

"When Dr. Jane looked at me first, she looked at me and just loved me. Ask Dr. Jane even if we're not together. She knows that old man Almas was a good researcher. And the book, she gave me a book. She wrote, 'Thank you, Almas, for your research. You have been very helpful to many students and to many of your colleagues.' And if it wasn't for all this chimpanzee stuff, I wouldn't have been in the forest." Nor, he adds, would he have climbed the mountains and learned all about the herbs and the plants that grow there. "That was taught to me by my mentors and by the chimps."

Among those mentors, Mzee Yahaya mentions another key figure from Gombe, Dr. Shadrack Kamenya. "How do you start building a house?" he asks, rhetorically. "You cannot build a house without a foundation. Dr. Jane and Dr. Shadrack started teaching me. So, they were my foundation." In fact, Dr. Shadrack Kamenya and his colleague, Dr. Deus Mjungu, not only helped provide a solid foundation for the work of JGI but are still there today, decades later, pillars of support, leading the new generation of field researchers from the nearby villages—researchers who continue the longest chimpanzee study in the world while also serving as chimp ambassadors and a bridge between Gombe and neighboring communities.

Chapter 8

A confluence of disciplines

Shadrack Kamenya explains why indigenous researchers are essential to outreach efforts

Deus Mjungu dedicates his career to creating habitat corridors for endangered wildlife

Behind the team of local field researchers in Gombe stand two exceptional individuals: the director of conservation sciences and the director of the Gombe Stream Research Center, positions that require a specific combination of local knowledge and academic accomplishment. Living most of the year in Gombe and Kigoma, the directors are consumed by their positions, never really able to leave their day at the office behind them. With the complex social dynamic of live-in staff including the research assistants, university professors and their students, visiting tourists, and government officials from TANAPA, the center is a hive of multiple cross-cutting teams, with a web of coordination requirements. Added to this is the ever-present pressure of maintaining and advancing one of the world's most significant scientific legacies. For many years, these large shoes of leadership have been filled by two cool-headed people, Dr. Shadrack Kamenya and Dr. Deus Mjungu.

A western Tanzanian local, Shadrack almost always appears as though he has just returned from an extended vacation, with a relaxed countenance and a casual and quiet manner of speaking. Despite his laid-back

Dr. Shadrack Kamenya (*left*) and Dr. Deus Mjungu. *The Jane Goodall Institute.*

demeanor, Shadrack never seems to stop working, acknowledging, with a laugh, "I'm actually retired, well...I'm supposed to be retired."

Even though he was born about 25 miles northeast of Gombe, Shadrack had never been there until a brief visit in 1990. "When I visited looking for an area to do my dissertation work," he explains, "I met several field assistants who were working with Dr. Jane. Looking at them, looking at their relationship with Jane and other researchers, I knew they were locals, bringing important experience from the villages. In December of 1994, I returned to do my fieldwork on red colobus monkeys, studying their ranging and feeding behaviors. Tacare was young then, only a few months old, and I began to learn more about what was happening outside the park, like their revegetation initiative. I liked that JGI were also engaged with HIV/AIDS education, trying to stop transmission, which was important for the community.

"In 1997, I became the director of chimpanzee research, by which time you could see that the situation in western Tanzania, particularly within the Kigoma district, was dire. If you took a boat from Kigoma to

A red colobus monkey on a rocky shoreline. *The Jane Goodall Institute. Bill Wallauer.*

Gombe, you could see the desperate situation of the villages—they were bare, no trees around their houses or across the hills, causing problems of soil erosion."

Now, Shadrack says, if you want to see the results of Tacare's outreach, "you just take a boat to Kagunga, near the border with Burundi, or you go to Kigoma. You look at the houses, you see all the houses are surrounded by trees, and that is a very good outcome of the work. If you look at some of the villages here—Kigalye, Mtanga, and others—you find the forest coming back in those areas."

When Shadrack works in the villages around his birthplace, he is among his people, so, he says, "they recognize I'm invested in them, which helps to gain their approval and buy-in to our outreach and ideas. I have seen firsthand the enormous power in having local indigenous people and even other Tanzanian nationals working with us. I have seen this be a particular advantage for the institute when engaging people on some of the more sensitive subjects of change."

Even so, he admits, being an indigenous person doesn't automatically guarantee trust from the local people, and, with culturally sensitive topics such as family planning, the discussions can be difficult. "They take time,"

Shadrack says, but the duration of the project has been helpful in this regard. "Now that JGI has been working with a number of these villages for a long time, the community members are likely to trust not only project staff local to the area, but also those from outside the Kigoma district, even international staff.

"When we take on subjects that are hard to understand," says Shadrack, "for example promoting alternative practices for land use, you are promoting conservation in the village forest lands to which they are directly linked and reliant upon. I stand at a point of saying, 'Look, this is what we are doing and why. We are not doing it because we want [your land]. It is because we want to protect this forest for the watershed and surface water retention of streams and rivers; we want to protect it for the wildlife you have, which is yours to treasure. We protect it because it is especially important for local climate, for climate change and for future stability. It further helps the process when we add our focus to other activities where direct benefits are easier to imagine, like supporting agroforestry or practices in my homeland like coffee farming and coffee bean promotions, for example. This also allows me to use the opportunity to educate my own family. Even though my native status makes it easier, trust is still always built with time, and it is a continuous process."

From 2003 to 2004, Shadrack was engaged in the Gombe Research Education Program (GREP), in which staff from both Tacare and TANAPA delivered outreach and education by visiting villages together. "Outreach is the only way of changing the perception of what we're doing, and whenever we've tackled that obstacle, we see people realizing that what we are doing is more beneficial than they imagined," he says, noting that reaching enough people is an ongoing challenge. "Environmental and conservation education in communities can be powerful, and wherever we do have the resources to reach, it is well received."

Shadrack can recall numerous instances of people either approaching him or talking to others about their experiences with Tacare and how it has benefited them. For example, he says, there were women who were giving birth every two years, "and all these children prevented them from other activities. When we started outreach on the topic of family planning, local people realized the need to plan their births in terms of frequency and space out childbirth in order to regain their health in between, as well as using their time for other productive activities like farming or operating a small business. I have heard a number of women talking about this freedom.

"Also, I've heard a number of people, one of which was from the villages along the lake in Zashe, who were among the first to plant trees for harvest, and during the last five years, they've been harvesting, cutting trees for timber, for housing, for furniture and selling, showing the sustainable benefits of even the long-term involvement. I remember one guy in Kalinzi who was a village nursery attendant before he became a village forest monitor. He continues to use the skills he gained as a nursery attendant and even now, still keeps his nursery for seedlings, which he sells. The money earned now pays to educate his children. You'll find these kinds of stories in various places, and that is very impressive. What's more is that they transfer this knowledge to others—all the villages in the Kigoma district, most of them are part of the project area. The villages in Uvinza, all of them are part of the Tacare project area. Now, in these villages, although we went in to promote tree planting, agroforestry, you find people who copied the ideas and techniques of village forest reserves, and some decided to set aside their farms and allocate them or part of them towards forest reserves."

People like Shadrack are good examples of how the confluence of being local to the area and having an academic background helps to keep the tributaries of understanding and relevance flowing though his work. When other streams of participants, like the government officials of TANAPA, merge into the flow of outreach, they can provide additional support otherwise beyond the means of non-government organizations like JGI. Yet, given the often-remote location of East African villages, anyone not born and raised in the immediate area is generally considered an

Dr. Deus Mjungu following chimpanzees in Gombe. *The Jane Goodall Institute, Nick Riley.*

outsider—even a compatriot. One such Tanzanian "outsider" is Dr. Deus Mjungu.

Reflecting on his early interest in conservation, Deus says, "I was lucky because the place where I grew up wasn't far from the Serengeti National Park, so when I talked about helping wild animals, the people around me knew at least what I was talking about and why. Slowly they began to remember about our own childhood, when we used to see lots of animals, and over time, the number of animals reduced, and the distances traveled to find them increased. I wanted to study conservation because I didn't want our own children to have to travel even further to see these wild animals."

Deus has spent more than two decades working in Gombe, arriving in 2001 after earning a zoology degree at the University of Dar es Salaam. He first worked as a research assistant under Professor Mike Wilson from the University of Minnesota (UMN), who was researching chimpanzee intergroup aggression at the time. Deus stayed for over a year, before leaving to do his PhD at UMN, during which time he would travel back and forth to Gombe. Since earning his PhD in 2010, he has been working as the director of the Gombe Stream Research Center, managing the center and overseeing day-to-day activities such as data collection, staff coordination, and other administrative duties.

"If I were to compare today with my arrival back in 2001, I would have to say that the local people are beginning to understand the importance of conserving the Gombe region and its wildlife," Deus reports. "Due to earlier land use practices, the people were really feeling the pinch of destroying their environment in terms of resource availability and have since seen the more recent benefits of altering their behaviors."

Deus explains that, although he doesn't currently carry out Tacare work for JGI, he is a stakeholder in the approach, as are all those living in the area. "There is a great need for having connectivity for the chimps because Gombe is too small and isolated, and without connecting Gombe chimps with other populations, Gombe is doomed. Connecting forest between Gombe and other chimp habitat like Burundi to the north is critical for allowing immigration and emigration of chimpanzee individuals to keep avenues of genetic transfer open, and the population healthy and viable." Creating habitat corridors like the one Deus mentions is a method widely adopted by conservation practitioners because it could benefit other species, and connecting one protected area with another through privately owned lands can benefit all biodiversity in a particular region. However, it is a difficult goal to achieve, as the practice almost always involves vast numbers of landowners and land titles held by various entities.

"They have been doing this by using Participatory Village Land Use Planning," Deus explains. "Now we can see some forest beginning to come back north of Gombe. Of course, there are some challenges because not all people are on the same level in terms of conserving forests outside Gombe, so JGI needs to continue to work with them regarding the need

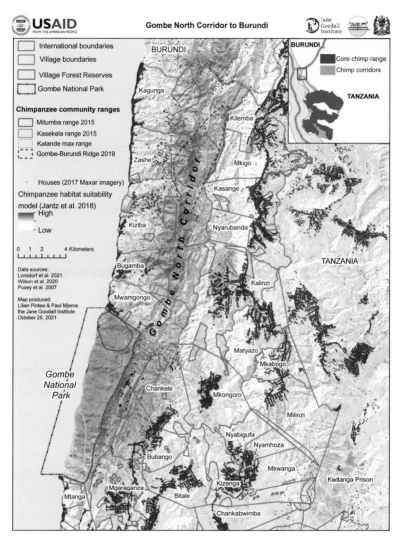

A map of the Gombe north corridor comparing areas of chimp populations in the greater Gombe ecosystem with human housing. The map was compiled from habitat suitability modeling, Landsat and Maxar satellite imagery, village boundaries, and forest reserves. *The Jane Goodall Institute, Lilian Pintea.*

for saving those sections of the corridor still at risk. I am sure those people won't understand this need unless we are addressing their major concerns by trying to reduce their poverty, making sure their basic needs are met, like having food on the table. By trying to provide alternative livelihoods, JGI has been trying to boost their economies with microcredit initiatives for small businesses and implementing Community Conservation Banks (COCOBA). Conservation should go hand in hand with helping people meet their daily needs, and if we can achieve that, it will help the people justify conserving the forests they have agreed to set aside." Deus adds that, at the research center, they try to recruit locally, which also helps address the level of poverty around the protected area.

"Addressing the needs of the people as a focal point makes more sense than just planting trees and being focused on a single species or habitat. These surrounding communities are so complex, and with their problems, you cannot just focus on one. Because if you do, say, community health, it fails to address other issues like access to water, schooling, and livelihood—to the point that it will even undo the work on community health that you are trying to achieve. So, I believe the holistic approach is the best way. Resources are not abundant, and if you can use the few resources you have to address multiple underlying issues, then that's both more beneficial and more sustainable. It means no set of circumstances are the same, and NGOs like JGI should continue to embrace the need for adaptive management and flexibility. It's a dynamic approach, not a static one."

Living inside the human-made island of Gombe, Deus knows that the future of protected areas like the one where he manages long-term research relies as much on the efforts of the people outside the park as of those inside, and habitat connectivity is the only real future for small areas of protected land like Gombe. Likewise, as director of conservation sciences, Shadrack understands that the future of a biodiverse region like western Tanzania depends on the custodians of unprotected lands. After all, counting chimpanzees for scientific surveys is simply documenting extinction if the cause of decline is not also thoroughly addressed.

"For me as an indigenous person," Shadrack says, "I know that we have several people who have learned a whole lot from the Tacare principles and so it makes me hopeful that the locals we have managed to connect with are likely to carry forward some of these activities. They may not connect the entire holistic package altogether, but you may find somebody doing agroforestry, somebody is protecting the watersheds, somebody doing family planning, and so on. I'm sure the experiences won't die out; instead I think they will continue to spread. Today, if you look at people's houses even in Kigoma, most of them are surrounded by trees, which is something that people are learning from one another." For Tacare's future effectiveness, Shadrack again emphasizes the importance of outreach. "A way to reach more people with the proven methods, proven approaches, that is the key. I know we are reaching people, but maybe not as many as we could be."

Even a mighty river such as the Congo is fed by numerous small tributaries whose combined forces build into what eventually becomes one of Africa's largest river systems. Similarly, environmental and social development programs are built upon numerous streams of separate disciplines, all trickling in from headwaters such as academia, local knowledge, and national policy. Yet every river has a source, or sources—points of origin without which its waters cannot keep flowing. From such humble beginnings, powerful forces can arise. And that is why Jane Goodall started the Roots & Shoots program.

Chapter 9

The cycles of old and new

Japhet Mwanang'ombe educates and inspires the younger generation

Hamisi Matama preserves the traditional ecological knowledge his mother taught him

The tropical forests are among the most biologically diverse and significant biomes in the world. A rain forest, for example, is so ecologically specialized that it creates its own localized climate, which it needs for self-preservation. Essentially, rain forests generate their own rainfall: the dense vegetation traps massive amounts of humid air that condenses into rain clouds, which, in turn, sustain the vegetation. This self-reliance exemplifies one of the core processes in nature—the positive feedback cycle. Both positive and negative feedback cycles govern how the world functions, and every natural process can be traced back to one of these two cycles.

The same can be said for human behavior, at least metaphorically. When a person initiates positive change that directly benefits another, the beneficiary is likely to feed this back through appreciation, growth, understanding, or even action. If the feedback cycle continues, affecting not just one individual but others around them, it gradually spirals until, one day, a self-sustaining habitat of community change and development has emerged.

Japhet Mwanang'ombe (*left*) and Mzaa Hamisi Matama. *The Jane Goodall Institute, Adam Bean.*

Japhet Mwanang'ombe, currently the national coordinator for the Roots & Shoots program in Tanzania, is a walking example of the positive feedback cycle. Born along the shores of Lake Tanganyika in the southern part of Kigoma, Japhet is a self-described man from the Rift Valley. "As a grade seven student of Bubango Primary School, I joined Roots & Shoots in 1991, after Dr. Jane, Kikwale, and Anton visited my school to speak to the pupils and teachers about environmental issues," says Japhet, brimming with youthful enthusiasm. Clearly energized by the subject of wildlife and environmental conservation, he became a Roots & Shoots volunteer in secondary school, and by 1998, had earned the position of coordinator for the program in Kigoma. "I grew up in the landscape and was part of the people who had impacted the local biodiversity in a negative way. Now, as a professional, I feel like I must pay the price. One of my goals is to help recover what we've lost in the landscape, after now understanding the impact of our actions."

Going on from secondary school to achieve an undergraduate degree in wildlife conservation and a postgraduate degree in natural resources assessment and management, Japhet represents how powerful the positive feedback cycle can be when young people are inspired. "When I grew up as a child," he explains, "I was so passionate about agriculture because I came from a farming family. When I met Jane and when I started with Roots & Shoots, it changed my way of thinking about our resources. I was more inspired by crops and the land, the health of the soil, the microorganisms. Before this, I never thought of the forest and its connections with rain, moderating climate, or as a habitat for fauna. I never actually thought of wildlife at all, particularly the chimpanzees and their need for an ecosystem with connected dispersal areas like forested corridors."

Japhet goes on to list all the other environmental issues he'd never thought about while growing up: "I never thought of the common resource-use conflict between the people and the wildlife, resources like water. I never thought of human encroachment as an issue, agriculture as an issue, the demography or population increase as an issue. I also didn't think about how socioeconomic activities could be positive or negative. When I was inspired in conservation, these are all the aspects I began to realize as critical for sustainability for both humans and the environment. When I studied wildlife, I was largely focused on zoology, looking at the behavior, looking at the microbiology of the animals and their health, anatomy, and physiology. Then I started to learn that it was more than that. It was more of an issue of ecosystem management," looking at conservation policies but also focusing on economic activities that are more compatible with wildlife conservation.

Through Roots & Shoots, Japhet says, his role has been to inspire and teach, to educate the young generation that they are in control of their own natural resources. "Time is not on their side for complaining about the mistakes and the problems their parents are creating," he points out. "For its size, Tanzania is one of the leading countries for deforestation, and in some regions more than 1,000 hectares of forest is gone every single day. We cannot complain while sitting in a meeting. The young people must understand that they are also part of it."

The Roots & Shoots program has visited every single district in Tanzania, Japhet says, and has more than 3,000 groups around the country

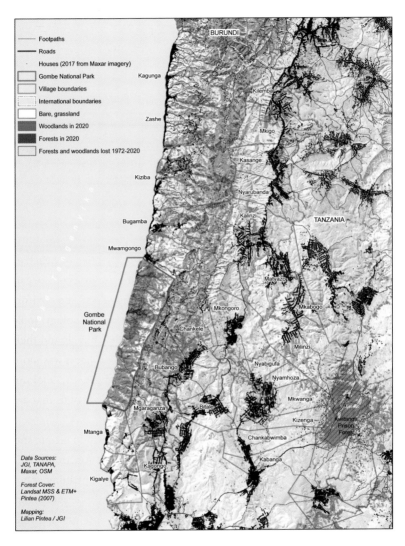

The extent of forest and woodland loss primarily outside Gombe National Park from 1972 to 2020 (pink) along with the locations of human development. *The Jane Goodall Institute, Lilian Pintea.*

Roots & Shoots students at the Sokoine School near Kigoma, Tanzania.
The Jane Goodall Institute, Shawn Sweeney.

and a database of half a million registered members. The program also has numerous social media platforms, connecting young people with interactive lessons and activities. As a coordinator, Japhet is responsible for producing educational materials for schools, education offices, the local government, and also the central government, since the goal is to get everyone involved in conservation movements and empower people to take action.

"We design different projects under the Tacare approach and try to work with the village government to provide the technical advice on how they can live in a compatible way with their surrounding landscape. We introduce them to native, user-friendly species that perform well as substitute fuel species for firewood, charcoal, and timber. Whenever they want to do agroforestry, for example, they are able to get to the species that are conservation-friendly, and we also provide watershed protection species."

In addition, Roots & Shoots works closely with the village environmental committees and school committees, focusing on the parents. These

Roots & Shoots students in Tanzania showing a treetop beehive on their school grounds. *The Jane Goodall Institute, Shawn Sweeney.*

committees help build the villagers' understanding as they participate in discussions about the historical trend of their resources, "how it used to be in the past, how the situation is now, what they want for the future. Then we consider what it would take for them to achieve a future they want," he says.

A key approach to Roots & Shoots classes is hands-on participation, which presents the information outside of the imaginary or abstract space and makes each lesson physically real and relevant. This helps students understand the connectedness of humans, animals, and the environment and solidifies their impact when they take action.

Like Shadrack, Japhet emphasizes the advantage of being a local, native to Kigoma. "You used to be a child in the same area," he says. "Your history is still there. It is easier to get trust from the community, and it helps the project become accepted in the village because then they see that somebody from their own village, somebody from 'our' landscape, is also part of the project.

"These are the village lands, and no matter how, they have to be compatible with both livelihoods and conservation. Conservation farming was one of the pilot approaches to help people, rather than evicting people from the landscape. The best way was to try to use innovative methods to make the same land compatible, and the best way to do it is through conservation agriculture as a method. Some call it smart agriculture and others called it agroforestry. What we say as a general term is *conservation agriculture*."

One of the techniques used in conservation agriculture is contour farming. "Farming on the valley's steep slopes was an issue, but creating terracing for the agricultural fields helps erosion and creates a more productive yield with less effort. In contour farming, the use of grasses to limit or to minimize erosion is one of the key aspects. The people in the villages, they loved that. We used a pilot plot for an example so the people could see it in action as a solution, but of course, it took a long time, and the project ran out of funding."

Another limitation, according to Japhet, was that he and his staff never had enough time to teach people about conservation agriculture. To implement it correctly, there are certain considerations: "If you don't have enough expertise, for example, you might end up promoting invasive species in the areas. One of the aspects was to try to help people understand what the best species are for use in the villages." Sadly, Japhet says, that project also lost funding and could not continue.

Despite these setbacks, he is proud to point to the project's successes. One such success was in Mwamgongo, where the source of the local water supply is a waterfall. "This waterfall was protected using the conservation approaches, by which I mean the people from the soil and forest conservation and community development department, they were working together, trying to empower the village committee to come up with some sort of a bylaw. One of the principles in the bylaw was to try to implement a 60-meter farming exclusion area around the waterfall, rather than people farming too close and destroying their water source. That success was the result of Tacare sensitization."

Mzee Yahaya Omary Mawisa, a Tacare agroforestry pioneer of the Kanyovu Cooperative Union, showing a coffee nursery and demonstration plot in Kalinzi Village. *The Jane Goodall Institute, Shawn Sweeney.*

Citing another positive result, Japhet says, "If you are traveling from Kigoma going to the border of Burundi from Gombe, you can see in the villages there's a lot of forest cover behind the village. But in the village itself, there is a lot of the tree species Senna siamea. Although they're there now, in the past it was bare. Now, when you look at the satellite images, you can see the differences between the historical landscape and how it looks today." Using VLUPs, the Tacare program has been able to facilitate local communities in establishing these village forests and using joint Land Use Plans to connect multiple village forests into an elongated corridor, creating a buffer outside Gombe National Park. This, in turn, has allowed the local migration of different chimpanzee individuals.

"Through each phase, we were able to come up with some key lessons learned," he explains. "Sometimes you'd introduce an approach to test if it would work, and if so, what time frame can be expected. Or maybe you'd try something else, which could be the same intervention but maybe a different approach. That's why the issue of capacity-building of the village government was such a focal point. It has never been a one-way approach. It has been a dynamic approach, taking the lessons from one phase, then building these into the other phases. I'm actually glad that I was also part of those PRAs to understand what people value, but also what they understand, and what they want. Conducting a PRA is important before introducing any intervention. The participatory planning helps to integrate indigenous knowledge within the planning process, to nominate the priorities."

For Japhet, traditional knowledge of the ecosystem is a key consideration. "It's important because in Tanzania, the people are living in the villages according to the law, and they're using the law of the Local Government Act of 1983. According to that act, people are the custodians of their heritage land in those villages, so the land is theirs. We can coordinate a land use plan, but it is the villagers who are responsible for developing and designing land-use bylaws—it becomes their responsibility. Here, indigenous knowledge of the landscape is helping them to understand the value of the resources they have."

Japhet describes how, in the village of Kalinzi, different water sources had almost completely dried up over the past decade. This was because some people were planting eucalyptus. The eucalyptus was introduced to the area by the missionary churches because it was a fast-growing species and good for timber production. Yet, he explains, the people planted them "without understanding the ecosystem services principles and how eucalyptus removes all of the groundwater."

To address this problem, Tacare proposed planting another tree, *Khaya anthotheca* (East African mahogany), as an alternative species. During those discussions, Japhet says, "the people started to understand, 'OK, now we remember how it was before the eucalyptus, they [the trees] used to be like this. Our forefathers told us about this, and this, and this.'" He adds that, locally, "many of the species were used as medicinal plants before the introduction of the health centers. The people were using the same species, the same vegetation, but now generations are passing, and things are changing."

As Japhet understands, trying to maintain cultural heritage is a balancing act between the benefits of development and the value of tradition. Hamisi Matama is someone who still holds one of the keys linking traditional remedies and beliefs with modern knowledge. Now retired, in 1972 he was one of many local villagers who were employed as research assistants, guiding foreigners and collecting daily observational data on the Gombe chimps. Yet, as a medicine man, his local knowledge of the vegetation is not only scientific but cultural.

"My first source of knowledge on trees was my late mother, who knew a lot about them," Hamisi explains. "She would teach me lessons like, 'My son, this medicine is to make you vomit if you've taken poison.' I started from there and continued progressing through inquiring and finding out more. I enjoyed it very much because it gives you freedom of understanding. The knowledge was beneficial then, but I still use it—even here at home I have planted many trees. I am still assisting many people who come for help. But at night, some people come and steal the medicinal plants. They don't request it, which means it will not work, but if they come and request the plants, then it will be powerful. For those who steal, the plants become ineffective. When people steal or cut the trees, I would advise them about it. I would not like anyone to do anything destructive."

Unfortunately, as is so often the case, the youth of the communities are gradually ignoring the teachings of their elders, dismissing their cultural heritage in favor of skeptical modern attitudes. As with many other facets of the culture, reliance on traditional ways of living has declined as the encroachment of the modern world offers alternatives that have undermined this knowledge.

Hamisi explains, "I used to add on to my knowledge of trees while I was in Gombe—there are things that I learned. There are plants that like to grow there and have a smell, which long ago we used to call 'Back of the River.' I was taught, if you ingest poison in food and you crush this plant and drink it, the poison will be eliminated. I try to teach this while the children are still young. Others, if you tell them, they see it as nonsense. A long time ago, we were taught that if you see this plant, it has a drug, and if someone poisons you, we could take it. If you died from the poison, taking this drug would mean that that person follows you, they would also die. Then they would know they have sinned."

Culturally, this type of medicine is a component of traditional spiritual practices and is still a major aspect of culture in Tanzania and across Africa. It plays into almost all relationships, all considerations and decisions, and although the mechanisms for using either defensive or offensive spiritual practices have been somewhat modernized, this cultural way of knowing is one that NGOs need to consider. These variables, somewhat obscure and invisible to outsiders, still play a very large part in local society.

Japhet takes up the story from Hamisi and explains that some of the older generation are trying to remind the communities about the value of indigenous knowledge. "There is also a campaign where the retired Gombe field researchers are trying to stimulate casual conversations with the community members about the value of animals, that people in the past would use an animal species as a symbolic representation for a halo, or a totem. Similarly, trying to pass on information about the historical values of certain plants is also important because it's being lost. For example, if your wife doesn't get pregnant, you take her into the forest; there's a tree that she must take that helps. This knowledge means they maintain or increase the value of the forest. There are also cultural traditions from times past that should be remembered. For example, in some tribes, it was not allowed to cut certain trees. Women could only use the dry branches that have fallen. This indigenous knowledge helped to promote conservation."

For Japhet, it's clear where Tacare should be concentrating its efforts. The Tacare approach rests on five different pillars, he notes—community development, health, forestry, agriculture, and youth—but, no matter

Ana, an expert in building fuel-efficient rocket stoves. These slow-combustion wood-burning stoves reduce the amount of charcoal or firewood needed for cooking. She has helped install more than 150 of them for other families in her village, Kalinzi, Tanzania. *The Jane Goodall Institute, Shawn Sweeney.*

how integrated the model, there should be a key pillar. "If you take the Jane Goodall Institute, the first thing people outside will think is chimp research," he says. But for the Tacare model, Japhet believes forestry is key. Yet, at the same time, "with conservation in Africa, you cannot isolate agriculture, because it's the driver for loss of biodiversity. In practice, look at Madagascar, look at Congo—in all the mega biodiverse countries, agriculture is not separated from the conservation package. My advice is that we cannot divorce ourselves from agriculture. It must be compatible, but the question is how we want it to be done. For me, conservation agriculture needs to be the key pillar—it's an integral part of the region in terms of livelihood."

Japhet explains why he sees agriculture as a priority. "For example," he says, "if you are looking at microcredit, you can do Community

Conservation Banks (COCOBA), you can do education, you can do even village forest reserves. But the moment you drive your car back to town, these people are going to their farms. We need to look at their day-to-day livelihood and then tap into it from there, try to improve it and build on what they are doing. There are a number of models for conservation agriculture, and it's proven to work very well."

Returning to the topic of Roots & Shoots and its work in the schools, Japhet says, "I think we need to look at broadening that dimension. I can give an example of the village Zashe, one of the greater Gombe ecosystem villages. In 2002, we used to have Roots & Shoots groups in Zashe Primary School, so the children were learning about tree planting, the environment, etc. In 2012, 10 years later, the same children who were in school are now grown up and have their own families. They then began complaining to their parents. Their parents were engaged during the Tacare land use plans and put aside some of their farms as protected reserves, which was to connect the village forests for regeneration purposes. The parents who made those decisions during the planning period, their children have grown up and their family size has increased. Now, this next generation believes their parents chose wrong, and so they're claiming back the piece of land. This creates problems for resource use.

"I see the gap there, that the education we are giving is not integrating with what other departments are doing. My take is that the type of Roots & Shoots education within Tacare needs to be inspiring and teaching the practices of what the villagers and the parents are doing for the generation gap to be minimized. Rather than teaching about the importance of tree planting, it has to go along with understanding the soil chemistry, understanding the land-use practices, understanding about the other variables. The approach needs to be integrated."

Japhet's forward thinking shows why the Tacare approach needs to be fluid, adapting over time in response to trends that may emerge years later. As the Zashe example shows, constant re-evaluation is important to understand what the cascading effects are from one generation to the next.

The seeds planted in one generation may bear unexpected fruit in the next, as social and environmental conditions change. After all, in the natural world, feedback cycles can be both positive and negative.

Chapter 10

Seeking homeostasis

KANYACODA, VHTs, PFOs, KIKACODA: Working toward human and ecological health in Uganda

J ust as the health of an organism is dictated by its cellular function, a local community fails, survives, or thrives depending on whether the needs of its individuals are being met. This mirrors a state of balance in basic biology known as homeostasis—an organism's state of equilibrium, the self-regulation of its physical and chemical systems in response to a stimulus, maintaining optimum conditions for survival.

What, then, are the key mechanisms for reaching human and biodiversity homeostasis? Although the list is long, the essentials for most global communities include support, local governance, and participation. Yet, even with these basic requirements, what works or what is needed in a country such as the DRC may differ greatly for practitioners wanting to use Tacare principles in a country such as Bolivia. Even in countries as geographically similar as Tanzania and Uganda, there are variations and nuances to implementing the Tacare approach.

Across western Uganda, JGI has cast a wide net of engagement with local people living within or adjacent to chimpanzee habitat. JGI-Uganda also has a history of significant reach and sensitization with many agricultural landowners in between the isolated pockets of forest. This outreach has recruited an impressively diverse array of more than 1,800 smallholder forest owners organized in 15 private forest owners (PFOs) associations, who set aside or plant their own private forests to be patrolled by community-appointed forest monitors using smartphones and the ArcGIS Survey123 mobile app. At the same time, JGI has worked to facilitate the formation of village health teams (VHTs) in the region. The results show how local governance and participation are benefiting the districts of Masindi, Hoima, and Kibaale in western Uganda.

Dispersed throughout the Masindi district is the Kasenene Nyantonzi Community Development Association (KANYACODA), which encompasses the Siiba Community Development Association (SICODA) and various VHTs which, combined, make up a significant force of 170 individuals, 53 of whom are women. Unlike programs in western Tanzania, which are driven almost exclusively by male participants selected by their village governments, gender participation in western Uganda is more evenly weighted by the local communities. These various groups all volunteer their participation across a spectrum of community development activities, under the categories of health, clean water and sanitation, forestry, agriculture, livestock management, and sustainable livelihoods, the domino effects of which are reshaping their relationship with the ecosystem around them.

As a group of KANYACODA workers gathers for a discussion, the first to introduce himself is Jackson Varongo, who has been a member of the VHTs since 2010. "We travel house to house promoting the Water, Sanitation, and Hygiene (WASH) program," he explains. "Our aim is to mobilize people in our communities and teach them about preventing water- or sanitation-related illness and even treat some of these ourselves—diseases like malaria, diarrhea, and so on. Otherwise we link them to a health center. We have been fortunate to receive funding under SICODA to build water supply points. Although my village hasn't been able to benefit yet, I am happy many others in the Nyantonzi parish have had these water access points constructed."

Jackson says he's seen a great deal of change in how people understand the connection between water, hygiene, and illness, with a resulting improvement in health outcomes. "Now those in the villages know what to do, how they can keep their families clean and reduce the spread of disease, and this is very noticeable with how they're able to protect their youngest ones. Historically, they would also come for immunization, although we have run out of these recently," he says with an expression of mild annoyance.

"Besides immunizations," adds a fellow VHT member, using her first name, Harriet, "we tell people how to keep their house sanitary, to have latrines, and a designated rubbish pit. I also inform people to sleep under mosquito nets and to keep boiled water." Jackson adds that much of this knowledge and support has gone to the local schools, helping to create positive attitudes toward JGI.

This simple yet vital health information may be taken for granted in more developed parts of the world, so much so that it may be difficult to understand how basic hygiene and sanitation still need to be taught in some communities. Although many in the developing world haven't received the benefits of scientific or medical progress available elsewhere, remote communities used to enjoy healthier ecosystems, fewer diseases, and cultural knowledge such as traditional medicine. Although these still apply to some communities, for others the deadly combination of increasing population density, poverty, loss of traditional knowledge, overuse of natural resources, and a degraded ecosystem now leads to many preventable health issues.

Godfrey Angue, a stoic-looking man and a senior VHT member, picks up from Jackson's earlier point. "When we started in 2010, we were providing treatments for fever, measles, and other ailments. But since 2013, now seven years on, we have not had immunizations available in some villages, because the people number too many. As a VHT member in my village of Oganda, in Kasenda Parish, this is a struggle I am facing." What Godfrey and Jackson raise are valid concerns that demonstrate both the previous success of the VHTs and the constraints on local capacity as the rural population continues to increase.

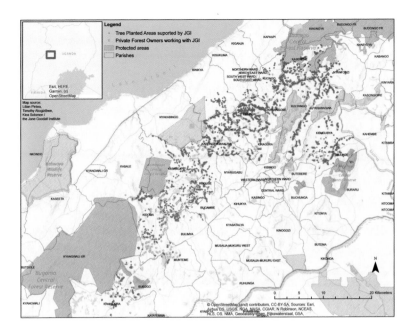

In this map of a section of Uganda to the southeast of Lake Albert, tree-planted areas, PFOs, protected areas, and parishes are shown. Various associations and VHTs work closely with JGI, using such maps to show progress and improve ecosystems. *The Jane Goodall Institute, Lilian Pintea.*

Zoonotic diseases are also an important consideration for those living adjacent to wildlife. History teaches us that bushmeat is a consistent source of disease transmission, with the deadliest viruses jumping from a vector species (animal carriers) to humans. Simian immunodeficiency virus (SIV) originated in West African primates, mutating into human immunodeficiency virus (HIV) after humans were subject to repeat exposure to the host animals carrying SIV, likely during bushmeat trade activities. Similarly, Ebola virus outbreaks occur through fluid contact between infected animals and humans, and in December 2019, the coronavirus disease (COVID-19) jumped from a vector species, likely a live animal being traded in China, and has since proven one of the most destructive pandemics in human history. Inadequate hygiene by people at the

Village health team, KANYACODA, Uganda. *The Jane Goodall Institute, Adam Bean.*

animal–human interface is part of the problem, but these and many other zoonotic diseases are primarily the result of our disrespect for nature and wildlife, with humans increasingly encroaching on natural habitats and expanding wildlife trade.

Proper prevention for livestock animals is therefore a significant consideration in programs where JGI provides founding stock for livelihood development in African communities. Godfrey Itigi, another VHT member, explains: "We've been taught many things about the healthy way to manage livestock, and how to prepare and handle their products. For example, we know not to eat undercooked goat, or raw milk products, because we get brucellosis bacteria from Brucella. Because of the information on zoonotic diseases we've received from JGI, we're eradicating these health issues. At first, the rate of deaths was high in this community but nowadays, our children are no longer dying as they were previously. This has also decreased congestion at the health centers, and for many preventions or treatments, the VHTs can assist."

Leading the livelihood development for KANYACODA and VHTs is JGI's Moses Kyalungonza, who manages the dispersal and management of village livestock, sustainable agriculture, and alternative income-generating initiatives. Together with Fred Tumsimi, who is the community member in charge of the KANYACODA group activities, Moses is guided by the principles of Tacare, which serve to reduce improper land use and lessen the local people's impact on the environment while promoting distance between humans and wildlife for a more balanced coexistence. "My role," explains Moses, "is making sure the communities are working as one for food security by promoting the enterprises that can be managed by the people from within their village. For example, we've placed over 240 beehives, each yielding about 5 kilos of honey annually.

"These beekeepers have been issued all the accessories to be successful," he continues, "including honey shoots, culture boxes, and harvesting equipment like a honey press machine. Regarding livestock, we have placed over 70 nanny [female] goats, and each community has a buck [male] goat to service the females. These were originally Boer goats, which are exotic, chosen because they grow faster and can be sold for a higher price compared to our local goats. The issue with these was because they're not from the region, they're susceptible to worms and can be expensive to treat. For this reason, we've chosen to cross the Boer goats with our local breeds, which gives them a higher resistance to parasites and maintains some of the Boer goats' attributes like weight gain."

"We have also transitioned improved varieties of cassava like NASE 14," KANYACODA leader Fred Tumsimi adds, "which is more suitable for the regional zone as this type has higher disease resistance. After each harvest, cuttings are taken and given to other people under the Pass on the Gift (POG) scheme. If someone's livestock reproduces, or their crops yield extra, these are passed on to another person who is without, and this way we increase the number of people that benefit from the association. We have 170 registered farmers and we're still registering. We are preparing other farmers with Napier, or elephant, grass and coriander, which are edible, palatable to goats and pigs. This way when they receive livestock as a POG investment, the locals know how to care for them, and the animals aren't mishandled. In addition, elementary practices like how to identify disease are passed on by upskilling certain individuals, who then go on to train other members in their village. We call them Teachers of Teachers (TOTs). These techniques take limited resources and time and help keep the development sustainable if or when we withdraw."

To support sustainable use of firewood, JGI has also designed and issued fuel-efficient stoves, which are simple slow-combustion stoves that both minimize the need for wood fuel and reduce dangerous smoke buildup inside the household huts. Again, these are distributed in limited number, while the technique for making more is passed on to the people in each community. All these enterprises go alongside the activities of KANYACODA forest monitors, whose data shows a steady increase in forest regeneration, with trees as tall as 30 feet taking just four years to grow thanks to the region's high rainfall. Such a large association of members has KANYACODA achieving promising results in health, livestock management, sustainable agriculture, and alternative income generation throughout the Masindi district.

Around 40 kilometers from the town of Masindi, the Hoima district's parish of Kibanjwa is home to a dedicated team of forest monitors, known locally as the Kibanjwa Private Forest Owners Association (KPFOA). This small but diverse team is driven to protect privately owned forest patches from meeting a similar fate to the surrounding landscape, which shows the heavy scars of agriculture and deforestation. This group introduces themselves by their first names only: Peter, Xavier, Oliver, Specioza, and Godfrey. As Peter leads the team down to a strip of riverine habitat, he proudly states that he's the owner of this 17.3-hectare forest corridor, which offers sightings of chimpanzees, colobus monkeys, numerous bird species, and more.

Xavier is the chairperson of the association and has been involved since it began almost a decade ago. "We started these activities in 2011,"

Members of the Kibanjwa Private Forest Owners Association (KPFOA).
The Jane Goodall Institute, Adam Bean.

he explains, "when JGI came and sensitized us about the importance of conserving the environment and keeping these private forests. We started monitoring these forests on foot but eventually after one year, JGI came out with technology which we used, called ODK, a mobile app running on Android tablets. That ODK tool, when combined with another app called Forest Watcher, assisted us a great deal in forest monitoring, especially when looking for the recently deforested places and any other illegal activities." ODK is an open-source data-collection software created by developers at the University of Washington that allows the forest monitors to use their Android mobile devices, smartphones, or tablets to mark, georeference, and share their activities and observations as GPS points. Building on the success and lessons learned from using ODK, forest monitors are increasingly switching to Survey123, which also allows them to visualize and communicate their data as dashboards and maps, using ArcGIS Online in real time. These conservationists are still working on foot, but a suite of mobile data-collecting and cloud technologies, facilitated by JGI within the broader Tacare framework, has allowed them to concentrate

KPFOA staff head into the field to check on a forest patch under their protection. *The Jane Goodall Institute. Adam Bean.*

their efforts and record their field data for later analysis and use by local, regional, and national stakeholders.

Xavier is the most vocal of the group, and it's clear from his outgoing personality that he takes pride in his associates and their mission. "Previously, I worked for the Uganda Wildlife Authority (UWA)," he says. "From there I learned of the benefits the forest provides. My fellow forest monitors and myself have been experiencing a lot of problems due to prolonged..." A brief pause.

"Drought," interjects Peter.

"Drought," continues Xavier, "and forest owners would plant their food, and, in the end, they would get nothing because of the drought. We came to realize that we had tampered with the environment, getting these problems, so people decided to say, 'Let us now conserve and protect ourselves so that we can bring back the environment.' That was actually the reason we are really conserving these forests."

"No," a quiet voice interjects. Oliver, a woman small in stature, politely begs to differ. "I joined this program because I saw that it was becoming a problem to get firewood in the village. In fact, even today people have reduced the number of meals just because of firewood. The firewood was not bought from the market but just picked from the forest or the trees around us. Now it has disappeared, and I saw it as especially important to have these forests recovered." Understandably, different people with their various roles in the household or community relate to the same issues from their own perspectives. As collecting firewood is an extension of cooking, women are often the family members responsible for gathering this resource, which becomes a time-consuming task as more of the surrounding forest is lost.

"But yes, water was the other thing," Oliver continues, "wells for the water. Water sources were drying up, so we were getting the problems associated with such a shortage. Through the sensitization, we came to understand that the water is reliant on the trees, because they draw the groundwater upward and keep it near the surface. With most of the village trees cut down, the riverine areas dried up. So, we came together to teach other people about protecting the wells by leaving the nearby large trees as exclusion…" Another brief pause.

"Zones," Peter offers, to the rescue again with a single word.

"Yes, zones," she continues. "I joined in September 2019 because I realized firewood and water were getting really difficult to find."

The tools used by the forest monitors began to evolve, making them more efficient and productive in their fieldwork. "After one year," Xavier picks up where he left off, "JGI came again with another tool, which was called the Forest Watcher mobile app. That one was even better for assisting us because, in fact, it was installed in our phones and it would take you directly to where the deforestation was." The smart technology behind the Forest Watcher app—originally conceptualized and developed by JGI, World Resources Institute (WRI), and Google Earth Outreach—uses Global Land Analysis and Discovery (GLAD) weekly deforestation alerts from satellite images processed by the University of Maryland and available through Global Forest Watch (GFW) to identify newer deforestation activity. The phone then alerts the user to the location, and the forest monitors can respond. "These alerts show you the deforested places," Xavier explains. "After opening, you touch on that alert, then that alert brings you a compass and that takes you to where the deforestation or any other illegal activities recently happened. You then touch the screen, and it gives you a form to fill out for collecting the specific information. Sometimes we move in a group, or sometimes we move individually—it depends on how we organize ourselves because we are covering all forest monitoring throughout the whole parish. In this parish, we have 68 active private forest owners."

Specioza, the other woman in this group of PFOs, seems to have come to the association with a naturally curious mind about the local ecology. Finding a brief pause in the voices of others, she says, "I chose to become a forest monitor because we are lacking many things. Our children and our grandchildren were not going to see the forest. Even when they study about it, they still might not be able to see it. I decided to work with this group, by planting and growing more native trees for the future. Some trees are medicinal so we sensitize people to reduce their pressure on the habitat so we can get herbs, which we use as medicines. Also, some animals were running up into our homes, destroying things because we have destroyed their habitat."

"I decided to join for that same reason," says Godfrey, catching up with the group after a detour by the stream in Peter's forest. "It's because the animals were coming into our homes. That's how I decided to join this group, to protect our natural forest. Even the idea of bringing new plant species so we could support these natural trees, adding on to where the villagers have cut the forest—I like that JGI did that. The animals that were coming into my home were baboons mainly. I saw it was good to join and sensitize other community members on how to keep the remaining forest so the animals had somewhere to live, and to show them how we can all benefit it if we…"

"Conserve," Peter says, not even waiting for a pause this time.

"What many of our community members forget," adds Xavier, "is that we're culturally connected here; we're from this kingdom. So, we came up with a suggestion that whenever someone dies, we would commemorate them by planting a native tree. Every person wishes to come and put something on the coffin but instead of flowers we put a [sapling] tree, then you take it and plant it in remembrance of the deceased. Once that started, we could see it was going well, so now we also do it for births. When you produce a child, a boy, you plant two trees, and when you produce a girl, you plant five trees. The local government are coming up with similar practices and are encouraging us to plant exotic tree species. With exotics, the trees grow fast and provide for the cultivation of firewood while giving the native ones time to grow."

It's worth noting at this point that the use of exotic tree species isn't always the best approach, even in privately owned wood plots. Australian *Eucalyptus* species were introduced into Uganda to satisfy the ever-growing need for construction timber, charcoal production, and firewood. As one drives throughout Uganda today, it appears as though these trees have taken hold both commercially and as wild growing stands. But as is often the case when introducing plant or animal species into an ecosystem that is not their own, the cascade effects can be troublesome, even catastrophic, for the natural balance.

In the case of *Eucalyptus*, now a hugely popular genus for commercial plantations, the trees are raising the economic status of Uganda and, at the local level, helping subsistence users with the shortage in timber. The problem is, the wrong *Eucalyptus* species in the wrong area can have negative ecological consequences, exacerbating the shortage of accessible ground water around riverine habitats. Most *Eucalyptus* species are drought-tolerant trees, designed to survive semi-arid soils, and all species use a taproot, a large vertical root that grows straight down deep into the soil, accessing the water table as it recedes deeper, like nature's own drinking straw. Indeed, eucalyptus trees were originally brought into Uganda, specifically Entebbe, in 1898 to soak up the wet, low-lying soils and shallow swamps to reduce the presence of mosquitoes.

Not only can they tap into these deeper water reserves and deplete them before they have a chance to rise, but the growth rate of eucalyptus in a moist climate like Uganda is extremely fast, making them thirsty plants. This combination of rapid growth and the longest "drinking straw" can devastate regional hydrology. When planted along a riverine habitat, for example, they can accelerate the drying of topsoil, while outcompeting what native vegetation remains; stealing space, sunlight, and nutrients; leaving village-adjacent creeks or streams all but dust; and threatening entire communities with dehydration. Although foresters, commercial growers, and NGOs are refining the most appropriate species or hybrids for their various needs, small-scale use like village harvest plots continues to be a critical focal point when providing these communities with saplings for planting, to ensure that the trees are not solving one problem while creating another.

Balancing the demand for firewood and for commercial, sustainable timber with the need for forest preservation isn't the only issue the volunteer forest monitors face in their daily duties. Roughly an hour's drive southwest of Xavier and the team in Kibanjwa, another team of monitors gathers from dispersed pockets of privately owned and protected forest patches surrounding Kidoma. These members of the Kidoma-Kabaale Community Development Association (KIKACODA) struggle with encroaching commercial crops and disgruntled peers wanting to continue land-use practices despite their long-term detrimental effects.

Nelson, an enthusiastic member of KIKACODA, is the most senior in terms of time served, having begun working as a forest monitor some nine years earlier. He is joined by Vincent, Jane, Fred, Ivan, Moses, Stuart, and Gertrude, all of whom cover eight separate villages across multiple parishes and subcounties.

"When I started," Nelson says, "the Jane Goodall Institute were helping to georeference the forests on private land. We georeferenced the

Members of the Kidoma-Kabaale Community Development Association (KIKACODA) group. *The Jane Goodall Institute, Adam Bean.*

whole parish and joined other parishes in Kitoba and Bulyango. Now I work in one village, that is Kyakatemba where we are standing. When JGI came and said, 'Work with us to identify the forest loss in your community,' I said, 'OK, this will help the forests come back.' When people understand that deforestation is bad, they'll slowly by slowly start leaving the trees, and I was motivated to help because our monitoring can act as a deterrent. We were using the tablets with the ODK mobile app and the GPS, marking the points of each site and each farmer in the association. That data was taken by JGI's Timothy [Akugizibwe], and he developed a map showing the forests in this area and the owners of the surrounding land."

Jane, a small yet bold and dignified middle-aged forest monitor from Butimba East, adds that the forests in that area were almost cleared. "After being made aware of the program, I wanted to join because I remember what these areas used to look like in years past. Now our villagers have

Nelson confirming and ground truthing a deforestation alert from satellite imagery using the Forest Watcher and ODK mobile apps. *The Jane Goodall Institute, Lilian Pintea.*

begun to plant 30 trees for every single tree they cut down. Although the demand is high, it's hard to meet. At first it was difficult to explain to people about the need to keep the trees, but one of the best ways of doing so is to explain how it can reduce the conflict between wildlife and people. It can reduce crop raiding, whereby chimps, baboons, and other monkeys ransack people's gardens and destroy the crops."

"Most human–wildlife conflicts," Nelson adds, "are near the big habitat areas between the forests of Wambabya and Bugoma." These two reserves serve as a pertinent example of how disconnected habitat creates islands of marooned animals, which inevitably reach human land-use areas when attempting to move from one forest to the other. These areas are often high-conflict areas, and the three kilometers separating the Wambabya forest from the much larger Bugoma forest to the southwest represent a perpetual checkmate on the checkered landscape of Hoima. "Animals from Bugoma try to move across," Nelson explains, "and chimps or other animals sometimes end up conflicting with humans. Because it's so open, it's usually not chimps because they come out, stop, and retreat to the forest. The common culprits on this stretch are the monkeys and baboons."

Fred, the forest monitor of Rwamusaga, adds, "When people plant many trees, animals from big forests like Wambabya, they come and stop at the edge of the forest where the crops begin. They used to destroy their crops like cassava, but when we started forest monitoring, we advised people to try and establish a buffer zone at the forest edge, between the trees and their crops. Although this works, it isn't popular because it uses up farmland. So, crop selection is important, as some crops are not eaten by primates."

Like Xavier in Kibanjwa to the northeast, these forest monitors use technologies for their patrols and data collection, including Android phones, mobile apps, and ArcGIS Online, facilitated by JGI, that can be customized to specific organizations or projects. Stuart, a forest monitor from Kihohoro, jumps to attention at the mention of the JGI-issued phones, whipping out his device like a sidearm. "We were provided with these Android phones," he says. "Whenever we're going to the field, there is the application we use that collects the data, whatever data we need. When you reach a place where someone has cut down the tree, we use this, it's called Survey123. It's very cool."

"When I reach the spot where I'm going to map," Nelson elaborates, "these forms appear. I select which form I'm going to use, usually it's the deforestation verification form, and I start filling it. In one place I first enter the date, the time, then I record location, then I start answering the questions like what is it? Is it crop farming? Is it livestock grazing? Large-scale logging? Massive charcoal burning? If it's for deforestation, how old is the deforestation? We just estimate based on how fresh or dry the wood is, and I capture the photo. This form has everything. Then I enter personal and location details and then press Save. I can either send it to the cloud or maybe keep in my phone."

Nelson explains that, after verifying the deforestation, the monitors ask the owner of that land why the tree was cut down. "He will give you reasons," Nelson says. "Maybe the tree was mature; it was old enough; he wanted to get money for school fees. Or if the farmer has cleared the forest, we tell him that it's not good to clear the forest. We advise that person to leave the trees and then we continue and report that incident to the Private Forest Owners Association. The association will also come and talk to that person who is destroying the forest. Then maybe that person will come to understand, but it's not always positive."

Stuart adds, "Some other people still think that they should cut the forest—they have a negative attitude that a forest can keep wild animals, which destroy their things or maybe harm their children when they go to fetch water. They want to clear it so that there are no monkeys or snakes nearby. For them, they think that cutting it chases the snakes away."

"In addition to that," says Jane, "sometimes people want money to keep them from cutting; others think we get paid to stop it happening. Usually, older people are easier to sensitize than younger ones because for the old ones, they remember what it used to look like, how much better it was. Now if we explain that by cutting the forest, the sunshine is too much or the rainfall is too little, that person understands. The younger ones will think of economic ways of getting money, saying, 'I want to cut this forest and replace with eucalyptus to get fast money.'"

Cash crops are another hurdle, as the promised income from larger companies is a tempting offer for many subsistence farmers. For instance, Moses from the Kikube district explains, "There is a factory which has been established within this community, and it's supporting communities to convert the land into sugar plantations. For the landowner, they do everything. They bring their machinery, they clear all the land for them, and the farmer waits for the time of harvesting, and he earns money. The factory does all the chore work of doing the planting, clearing the land. It motivates people to convert the land into the plantation. This is the challenge that we are facing now, and we have a program of engaging the sugar factory owners. That is still a process we are pursuing."

Another young man chimes in to explain their communities' approach to financial reimbursement and medical costs related to human–wildlife conflict, developed with JGI support. Ivan from Kidoma Parish says that KIKACODA has received resilience funds from JGI, and "now, that money is helping farmers mostly who are affected by human–wildlife conflicts, supporting them with that fund. These are loans, that they repay with 5 percent interest. 2 percent of this goes to the association, and the remaining 3 percent becomes insurance money in the resilience fund."

Ivan explains how the process works: "If a chimp crossed the buffer zone and then reached a farmer's plants and destroys the crop, we [the monitors] take that photo using Survey123; we record the location; we then take that information to the association. The association will also send members to go and verify. They also verify that the farmer has totally been disturbed by such and such an animal. Then the association will give that farmer some money, almost totaling to how much has been destroyed."

Gertrude, a young mother busy with her infant, adds, "Because we want them to benefit, we help them travel and receive treatment in a medical clinic. If a chimpanzee comes and attacks someone, the forest monitor goes to gather details, then reports to KIKACODA. Then we take the child or the one who is affected by the chimpanzee, and then that person is taken to the hospital and that resilience fund is how we pay for it."

This rather matter-of-fact discussion about attacks on villagers by neighboring chimpanzees reveals yet another obstacle when it comes to balancing conservation with community development. The topic is so contentious that it can make advocating for forests and wildlife more difficult and human–biodiversity homeostasis more challenging to achieve. It's not surprising that in some areas where the two species overlap, chimpanzees are persecuted because of the threat they can pose to human safety. After all, how would you cope with an attack on a loved one, especially a child? For the unfortunate few, sometimes living right at the edge of farmland and forest becomes a potentially fatal interface.

Chapter 11

The fatal interface

KACODA, Uganda: Finding successful strategies to reduce human–chimp conflict

As anyone who has personally spent time with her can attest, Jane Goodall has a truly engaging way of connecting with people. Indeed, she has even received criticism for maintaining her genuine kindness when dealing with individuals from industries well known for contributing to global degradation. Jane treats such people with respect and courtesy, connecting with them first on a human level. When she extends kindness, warmth, and a sincere willingness to listen, even potential antagonists engage in a dialogue that affords them dignity and respect.

That's not to say Jane doesn't take the opportunity to very politely remind them that their industrialized exploitation of earth's resources is degrading the planet, because she certainly does, albeit in a soft and even tone. But as Jane herself has said, "You can't bully people to change. You can't force conservation with an iron fist." It's no surprise, then, that she has maintained her position as a United Nations Messenger of Peace since 2002. Kindness, understanding, and empathy are not just words to her; she leads by example. The fact that "every individual matters," as she often says, applies not only to chimps or other animals but also to people. For Jane, there are always two sides, always cause to respect those who oppose, and often ways of bridging a divide by being open to viewing the world from the other's perspective.

Members of the Kasongoire Community and Development Association (KACODA). *The Jane Goodall Institute, Adam Bean.*

With this in mind, imagine a hypothetical global headline: "Three Chimpanzees Snared and Butchered by Local Villagers in Rural Africa!" Such a dramatic news item could easily conjure animosity toward the people responsible, even from those who are not passionate conservationists or animal advocates. But consider this perspective: A young mother living on the fringes of poverty and wilderness sits her child in the shade of a tree, just a few feet from where she is hand-plowing her vegetable plot. In response to a faint sound, she looks back just in time to see her child being carried off by a chimp, never to be seen alive again. The devastation any parent would suffer after experiencing this is painful to imagine. Yet this does occasionally happen.

A Land Cruiser full of globetrotting conservationists arriving in Africa to inform the locals they should conserve chimps for reasons A, B, or C would no doubt have a very short and unsuccessful visit. One cannot engender change by being solely focused on the symptoms of a problem.

There is no good or evil in scenarios such as the above, only desperate attempts by both animal and human to survive in a region where their required spaces overlap. So just how do communities live in these areas of conflict, and how can Tacare facilitate a system of coexistence through a mutual respect between the two species? The first half of the question is answered by JGI's outreach, education, and sensitization, the second half by the local people themselves.

In 2015, a group of locals from seven villages in Uganda formed the Kasongoire Community and Development Association (KACODA), which has worked with JGI ever since. As the association's chairperson, Henry Balemesa, explains, "The main mission behind the formation of the association was to address the community issues of chimpanzee–human conflict, which was increasing as a result of the surrounding forest cover disappearing." With two forest monitors in each of the seven villages, Livingston Muchwezi, the chief monitor, says, "We have worked with all the villages in Kasongoire and experienced both good and bad behaviors from the people and the chimpanzees."

As the 15 members of the association sit shoulder to shoulder in their headquarters, Henry tells the story from the beginning. "Right here around the community"—he points outside KACODA'S hut, beyond the congregation of curious children—"all the forest out there was finished. The decrease in trees forced the chimps to come out of the remaining forest into the communities in search of food," and the conflict escalated because the villagers were forced to collect water from inside the forest. "The water our community once relied on had dried up after the trees were gone. These people would have to collect water from the same place where chimps drink, and there were problems when people and chimps arrived for water at the same time of day."

Henry stalls for a moment, then continues. "Two small children lost their lives after they were taken by chimps high up into the trees. They were never recovered. A further five children were attacked but survived, although now they have scars and are still mentally disturbed." Henry's words are straightforward but piercing, especially with the group of eager young children gathered outside.

Funneled into a corner of dwindling resources, two large predatory animals seldom find harmony. "We have also had reports of other attacks occurring after villagers would chase chimps away from their crops," Henry explains, "throwing rocks and sticks to deter them. The chimps would return later that day and attack, in revenge—maybe. After all this, there was no peace in Kasongoire, and we looked for assistance in all other corners of government, but nothing resulted from this that could help the victims or their families. They still needed to collect water, though—can you imagine? This is when we as a community held a meeting and agreed that if we were to receive support, we would have to do so by forming a committee to lobby for assistance. This mission then became the formation of our association. When we lobbied, it was the Jane Goodall Institute that intervened," he says with a nod and a faint smile.

Henry explains that the first goal was to sensitize the community and teach the local people about chimpanzee behavior, because the most immediate issue was safety. Most community members didn't know how to behave in an environment with chimpanzees, he says: "We didn't even know chimpanzee behavioral habits or needs. How to avoid certain situations and how to respond to others was a priority. We were able to educate children through JGI's Roots & Shoots program, and schools around the parish were all sensitized and taught using illustrations and posters. The sensitization was generally done both at the parish level and the village level throughout 2015 and 2016. Since then, we have not recorded a single act of chimpanzee violence against a community member because people are now prepared with this knowledge."

Next, he says, hot spot areas were identified around water sources where children would fetch water. The children were taught how to identify water points that were too deep in the forest or surrounded by fruiting trees frequented by chimps. "With the help of the Jane Goodall Institute, we then received some boreholes outside the forest. This was a real positive point of the KACODA/JGI association because it provided us with alternatives to going into the forest for water."

Another hot spot area was people's gardens, where the chimps would look for food and raid the villagers' subsistence crops. The forest monitors now respond to reports of chimpanzee activity near water or gardens, then alert the community members to avoid the area. "I think this is a big part of the reduction in human–chimp conflict," Henry says. "But this still didn't solve the problem of chimps damaging subsistence crops."

To help solve this problem, the Ugandan Biodiversity Fund (UBF) and JGI helped KACODA start the Resilience Fund Project. This project employs two techniques. The first is identifying the types of crops chimps prefer, because many community members unknowingly plant those very crops. "Through more outreach, we introduced crops that don't encourage foraging raids," Henry explains. "That was mainly sunflower, soya beans, and chili."

The second technique was issuing compensation to people who had suffered losses during crop raids. This led to the formation of a human–wildlife conflict committee, which responds to reports of losses and issues a value of loss if the evidence shows chimps were responsible. "For this," Henry says, "we used grant money from UBF, which was 10 million Ugandan shillings [approximately $2,800 USD as of early 2022]. In order to make this grant sustainable, we decided to issue loans to all village land councils (VLCs) by dividing the grant money evenly and charging 2 percent interest on repayments. This 2 percent is kept aside from the 10 million [shillings] capital for damage or emergency costs, and this became a big success. We now have just over 17 million, and the additional 7 million is paid out to farmers for crop damage, or for medical treatment of injuries and cost associated with travel to a medical facility."

Henry adds that the human–wildlife conflict committee also needs to personally visit the location of each report because, he says, "we may not always pay for chimp damage in cases where the person has been taught to know better. Maybe they planted crops which are liked by chimpanzees around the forest edge and did so knowing the risks. If that's the case, we don't support them, because we sensitized and taught them, and they ignored the information. At times, we just give advice about the best crop selection for different areas and distances from the forest. For example, growing bananas too close to the forest edge is inviting financial losses, so bananas should only be grown in the villages further away from the forest. Soybean and sunflower, they are working better for those who are around the forest edge."

In addition, with the support of JGI, the villagers now have access to alternative sources of income, such as Boer goats, pigs, beehives, groundnuts, and various crop seeds. At times, Henry says, "if someone's garden is raided, he or she says, 'Instead of giving me hard cash, maybe you give me sunflower seeds. Maybe you support me with a Boer goat.' One Boer goat is worth 500,000 shillings. People make requests that better suit their needs. That's how we're coping with the situation." According to Henry, the people in the villages around Kasongoire have begun to change their attitudes toward chimpanzees; because the communities are being supported, they're beginning to see the benefits of maintaining the forests and the chimps around them.

Like their counterparts elsewhere in Uganda, Henry and his team practice Pass on the Gift (POG). Henry explains how it works: "If I'm given a Boer goat and it produces a young one, I pass it on, then it is even given to another person. Then if it produces again, they'll pass it to another person. If I get 50 kilograms of groundnuts and I harvest just 10 bags, I refund the original 50 kilograms back into the scheme so that it can be also given to another person to support another person. All the villages are participating in the POG. Our community is a proud one because I'm telling you, most of the communities around here are trying to copy what we were doing."

At this point, Johnson, a forest monitor, interjects to explain what prompted him to participate in the project—namely, his clan affiliation. "I'm a Musiita by clan," Johnson says, "and our totem is apes. What motivated me to join forest monitoring is because these apes are an endangered species. What I saw looking forward was that my totem might disappear in the future, so I joined as a way to protect this species."

"I can say this," Livingston adds, "to date we saved three chimpanzees: Gerald, Coco, and Saba Saba. They have names now. The first was Gerald in 2016. This chimpanzee had crossed a clearing for food, but in coming back, it fell in a mantrap. This is a mantrap," he says, holding up an aggressive-looking steel set of jaws. "You see, these are teeth of the mantrap, so if it is set like this, then a chimpanzee steps inside, it is caught. It's different from a snare. We saved Gerald through the partnership with the community. It was the community members that alerted us, as if our own brother was caught. When we responded we found him very stressed and moving with fear and violence. We had to alert the subcounty, Udongo, and reported the trapped chimp to the UWA and the Jane Goodall Institute.

"They responded in time and he survived, and his leg was saved. The second we rescued was Coco, in the forest on December 29 in 2017. She was caught by mantrap also. This one was a bit tricky because we didn't know she was trapped there at first. If they're with a group, they make their alarm call. It's easy to recognize. But with her, she was alone, so the typical loud group calls were absent. We estimated she stayed trapped by the hand for at least four days, so by the time she was found and rescued, her hand required amputation. Then last year in 2019, another mantrap caught a very young chimpanzee, that was Saba Saba. We had to remove her hand also. That effort was done by UWA because we have relationships with their vet services."

The change in perspective that the community of Kasongoire has undergone in five short years is nothing short of inspiring. From traumatizing examples of children's deaths and revenge attacks to sidestepping the human–chimp collision over land and resource use by improving subsistence agricultural practices, gaining capacity via alternative livelihoods, and developing a community-managed system of financial payouts for

A JGI-Uganda veterinary intervention team—led by veterinarians Dr. David Hyeroba, Dr. Peter Apell, and Dr. Tony Kidega—removes a jaw-trap from the wrist of a wild subadult female in the Rwensama forest in Uganda. The procedure was successful: She regained full use of her hands, and the chimp went back in the forest with the rest of her group. *The Jane Goodall Institute-Uganda.*

health care and property damage—the transformation is striking. Perhaps the highest level of respect shown by the KACODA team is that not only have these initiatives generated a truce of sorts between locals and chimps, but the humans also actively extend the kindness of rescuing trapped and maimed individuals.

A common thread throughout the Tacare approach is that contextual insight is critical. Under the long-term leadership of Dr. Peter Apell and his team, Tacare has been adopted in Uganda following similar principles as in Tanzania: engage, listen, understand, facilitate—and, for communities like Kasongoire, focus on the realities of the local people. As KACODA's success has shown, putting people's needs first, and directing resources accordingly, can be a way to ensure chimpanzee conservation as well.

Chapter 12

From the cloud to the ground

Uganda Wildlife Authority: Obed Kareebi, Frank Sarube, and Philemon Tumwebaze on poverty, technology, and conservation

Often, the boundary of a national park or conservation area is notional rather than ecological, a line on a map that is irrelevant to wildlife and wildfire, as well as to humans in search of subsistence. Disregarding this human-drawn line, wild animals will persist in pursuing their pre-existing corridors; humans living on the edge will continue to claim the resources they need from the forest. In Uganda, maintaining a sustainable equilibrium between the ecosystems on either side of that line is the task of the rangers and wardens of the UWA, equipped with the tools of their trade: tenacity, ingenuity, and technology.

Obed Kareebi is an ecological monitoring ranger for Kibale Conservation Area in western Uganda, which includes Kibale National Park, Semuliki National Park, Toro-Semliki Wildlife Reserve, and Katonga Wildlife Reserve. He has worked for UWA for seven years, responsible for wildlife health, waste management, and fire management, as well as ranger-based data monitoring. Joining him are his colleagues Sergeant Frank Sarube, the acting head ranger, in charge of security and law enforcement, and Philemon Tumwebaze, an assistant warden for ecological monitoring

Members of the UWA. *Left to right*: Frank Sarube, Obed Kareebi, and Philemon Tumwebaze. *The Jane Goodall Institute. Adam Bean.*

and research. Collectively, they represent years of experience in managing conservation issues such as deforestation, fires, and poaching.

According to Obed, most of the poaching in the protected forest is done with snares. "It's very easy for somebody to sneak in, lay a snare, wait for a day or two, go and check, maybe find a trapped animal and take it," he says. Staff from the UWA "tirelessly move to those areas," Obed says, "check around," and, with time, become skilled in tracking poachers and understanding their methods.

"We have animals in different areas," he explains. "Like there are some areas where you find a big concentration of buffaloes, duikers, bushbucks, elephants. Depending on the size of the snare, somebody [from UWA] will know this snare is meant for a specific animal." Assisting UWA in this task are other wildlife snare removal teams, some funded by JGI. "They have done a great job and as we talk, Kibale is one of the best-protected parks, partly because of having those teams who save the life of animals. The number of snares has reduced with increased effort, but this also goes

Wire and rope snares are placed by poachers on the forest floors, where the traps can catch and entangle chimpanzee hands and feet, causing grave injury or even death. JGI and partners have removed tens of thousands of the illegal snares in three forest blocks in Uganda. *The Jane Goodall Institute, Jen Croft.*

hand in hand with community sensitization. If communities outside the park have enough sensitization, some people now don't find any reason for poaching when they have [income-generating] projects to benefit from." Such projects are under way in these communities, Obed says, funded by UWA and its NGO partners, including Jane Goodall Institute chapters from Austria, Netherlands, Switzerland, the United States, and Uganda.

Frank adds that, in the Kibale area, UWA has formed associations for former poachers, who have been instrumental in removing snares. The idea goes back to 1997, when the Kibale Snare Removal Program was established by the Kibale Chimpanzee Project (KCP) in collaboration with UWA. Dr. Richard Wrangham, a professor at Harvard University and a former student of Jane's, set up the KCP long-term field site in

1987, inspired by his studies at Gombe. This project included the idea of engaging and working with local communities. With Elizabeth Ross, Dr. Wrangham cofounded the Kasiisi Project that, in the spirit of Tacare, has been supporting communities adjacent to Kibale National Park since 1997. Like branches from the same tree, these projects have been collaborating with Jane Goodall Institute staff across multiple chapters, supporting the work of local communities and UWA in stewarding their environment, wildlife, and livelihoods.

Frank explains how UWA works with former poachers: "In the first place, they themselves were poachers, so they know the areas where these wires are laid. What we did was to integrate them in our system so that they begin directing us to those points where these wild snares are laid. They direct us to those points, spot-on, and we begin uprooting like this." As an example, he describes a recent overnight patrol during which 85 snares were removed. Also, he says, the ex-poachers are acting almost as rangers, "because when their colleagues learn that they're involved in uprooting the wire snares, then they fear to go back because chances of arresting them are very high."

The maximum penalty for poaching or wildlife trafficking in Uganda is life imprisonment, Frank notes. "If you are caught with lion, elephants, those endangered species, if you are caught with ivory, for example: life imprisonment. If you are caught with other species like gorilla, chimpanzee—you go for life imprisonment." Sadly, he reports, there are very few, if any, lions left in the Kibale Conservation Area.

Philemon asks, rhetorically, "How do you keep ex-poachers motivated?" He explains that another form of poaching is prevalent in the parks—namely, pit traps. The poachers target elephant trails and dig long, deep pits, "where the elephants get trapped, so that the poachers can easily kill them and take the ivory." The pits are camouflaged with light brush or branches, "so when an elephant steps in, it just goes in and cannot come out. It's trapped. It can't move," and sometimes can't even breathe. "At times they [the poachers] put spears in those pits. I think that method is

A Uganda Wildlife Authority ranger crosses one of the elephant trenches in Kibale National Park, searching for potential deforestation sites and guided by satellite alerts shown in the Forest Watcher mobile app on his tablet. *The Jane Goodall Institute, Lilian Pintea.*

reducing now. Actually we aggressively addressed it using law enforcement methods."

In addition, Philemon says, UWA has launched a program of refilling all the pit traps in the park, recruiting ex-poachers as workers. When the reformed poachers are paid to fill the pits, he says, "another poacher who sees these people benefiting much and he's not benefiting from poaching, if he thinks twice, he also reforms and becomes part of this."

Yet, Frank points out, among the local people, there are cultural attachments to poaching, which, in some families, is an intergenerational practice. "It's something hard to break," he says, "unless maybe with enough education." Also, says Philemon, poachers tend to operate seasonally. "We have seen so many dry seasons. Communities here have tea plantations, so most of the people are engaged in tea plucking during the rainy season. They can't harvest enough during the dry season," so then "they get time to go to the park and set such a big number of snares. But also during festive season because people want meat."

Philemon adds, "It's not an easy job to do poaching. You can even go for two, three days," and return empty-handed. Noting "the time wasted, the number of days," he says, "If such a person gets productive work, I think he may get enough money to buy meat and get those proteins."

Obed agrees that most poachers "do it for subsistence survival," but he also notes the existence of poaching syndicates, spurred by the high prices offered by some hotels, where bushmeat is a delicacy, and by the market in China for pangolin and other African wildlife. UWA occasionally even encounters poachers who are armed. As in all syndicates, Frank says, the people on top benefit, while those on the ground assume the risk and earn relatively little.

Syndicates aside, Frank says, "Most of these people are poor. They do some of these things just to earn a living. We have both Kibale and Semuliki that are surrounded by big tea estates. These tea estates, they have casual workers. The kind of earning they get from this kind of work is also little. It is not enough to sustain them."

The problem is especially acute, adds Obed, because many local people have large families to support. "You find a person has five, seven kids. They need food; they need clothes; they need even school fees," which they don't have. When such a person is offered more than his monthly income for poaching a single animal, it's not easy to refuse. "That's why now we actually have that category of younger people joining just because of poverty," he explains. "Otherwise, if there was an alternative, they wouldn't be poaching. The unemployment rate in Uganda is very high and the youth population has really gone very high. That's the challenge that we have. We have to ensure that these animals are protected, but we have still that pressure of [the park] being an island among people who are poor."

That's why, Obed continues, UWA created its directorate of community conservation, to carry out conservation in partnership with the surrounding communities. "We ensure that we give these communities 20 percent of the revenue collected; we give it back to them to start livelihood projects. We directly give them the money through the local government arrangement. They write proposals; we assess the proposals; we fund the proposals. We fund the projects and keep monitoring them. We have given out many beehives to these communities."

In these community conservation projects, as well as its anti-poaching and reforestation efforts, UWA has been supported by technological tools and training from JGI, including mobile technologies and apps such as Forest Watcher, ODK, and recently ArcGIS Survey123 for data collection. Using the Spatial Monitoring and Reporting Tool (SMART) software and JGI-issued smartphones, Philemon explains, "We carry out routine analysis to tell management where illegal activities are concentrated, where wildlife are concentrated," and generate reports to guide deployment and planning. Frank adds that some of their staff are now able to use GIS for georeferencing and mapping, and, in turn, they train others. The Forest Watcher mobile app—developed by JGI, WRI, and Google Earth Outreach—was "a very important tool to us," Philemon says, "and we were lucky as Kibale to pioneer this tool in forest loss management. It helped us greatly to see how to work with the community because we used to have forest loss, and we could not tell."

After having adopted the Forest Watcher mobile app, Kibale was labeled as a hub of innovation, Philemon says, because the UWA monitors were using this tool so successfully. "It has helped us to control encroachment—very, very important to us," he says. Using the application on their smartphones to scan the forest, with alerts from space generated by the

Uganda Wildlife Authority rangers use weekly GFW and GLAD forest alerts and the Forest Watcher mobile app to locate and stop encroachment in Kibale National Park. This Maxar satellite image shows the park boundary line in yellow, GLAD forest loss alerts in pink polygons, and confirmed deforestation in red points, derived from Maxar satellite imagery. Kibale is to the left of the yellow boundary line. *The Jane Goodall Institute, Lilian Pintea and Timothy Akugizibwe.*

GLAD laboratory at the University of Maryland and NASA satellites, they were able to verify forest loss in two significant areas. "In Rweteera, we had lost over seven acres," Philemon says. "When we reached there, we found people had entered the forest. It is just a stone's throw from here, but we had not known about it." In another area, locals had encroached upon the protected forest and planted rice and maize. In both instances, Philemon and his staff were able to negotiate a peaceful withdrawal from the park; at the same time, they realized that the park's boundary needed to be more accurately delineated, a project for which they were able to secure funding from UWA. Thanks to the technology from Forest Watcher, he says, "we have not created conflict with communities, that we're taking their lands, because we verify live [with the software], and they see [the boundary]. Then we are living in harmony with them."

In addition to detecting forest loss caused by humans, the Forest Watcher app helps detect forest loss caused by animals. Philemon notes that human–wildlife conflict is an ongoing challenge for UWA. He explains how, during routine verification of alerts, he might see an alert in the middle of the forest. "You'd wonder, have people entered the park? You go there to verify; you find that most of the loss is caused by elephants. Also, in the northern part of this park, we have a lot of elephants that normally come out of the park and disturb people."

The reason, he says, is that there used to be wildlife corridors in that area, "so that's why you see elephants continue to force themselves to cross." Until recently, the elephants would cross there: "They were coming to Toro-Semliki, then they used the corridor that connects Kibale to Itwara Forest Reserve, then to Toro-Semliki, continuing to Semuliki National Park and Congo. But all those corridors have since been cut off." Not only elephants have been affected, Philemon adds: "We used to have lions, but they have not been seen for some time. In 2014, we had a lion that came. We didn't know where it came from, but those who were there before us tell us that lions used to cross from Queen Elizabeth [National Park] to Toro-Semliki and from Congo—something like that. But because of human population pressure those things have been cut off."

Having verified the concentration of elephants on the north side of Kibale National Park, UWA management has focused its efforts on controlling human–wildlife conflict there, mainly by constructing elephant-deterring trenches. "We have also established a whole unit with its headquarters in the northern parts of the park to ensure that every time elephants go out, they are pushed back in the park, because there are some areas where trenches haven't reached," Philemon explains. "Maybe if we had not worked with JGI to have that additional technology, community conflict would have intensified, and we wouldn't have had that initiative of putting in the trench and putting in that unit. Then also maybe encroachment would have continued. To me, that's how Forest Watcher, which JGI introduced here, has to some extent created a good relationship with UWA management and communities around us."

Another issue on which UWA is working to engage the local communities is reforestation. As Obed explains, part of Kibale National Park was encroached and settled in the 1970s; in the early 1990s, the government took action to evict these populations. Now the challenge is to restore those parts of the forest where trees had been cut down for so many years, mainly for firewood. Since 1994, UWA has been working with a Dutch NGO, Face the Future, on reforestation projects. "They have planted natural forests back," he says, but the demand for firewood in the surrounding communities continues to rise.

"To reduce that," Obed says, "the initiative was to have stoves that can save fuel. You save fuel so that the demand for firewood from the park can also reduce." Through UWA's partnerships with NGOs, "many families, many schools have been able to get those energy-saving stoves." However, he says, "you can't have every family having an energy-saving stove unless they are able to build it on their own, which we also train them to do. Sometimes they can't do it, so they require our expertise, but it's something we always do to make sure those who don't have [a stove] can get one. As the Uganda Wildlife Authority, we also plan to do it ourselves, as a demonstration. We have stations, outposts, all around the park. I think the next step is we are going to have energy-saving stoves in each outpost, because we also use firewood."

Reflecting on the technological tools that he and his colleagues now have at their disposal, Obed says, "I have to thank JGI, WRI, and the Global Forest Watch, particularly Timothy [Akugizibwe] for bringing that program here." With the Forest Watcher app, he says, "you are able to go through areas you have never been in, which have never been patrolled before. In the process, sometimes you discover something new as you navigate." Also, he says, the app helps UWA workers identify the areas where there have been fires, so that they can put their fire management practices into effect. For, as Frank says, philosophically, "What you don't expect today can happen tomorrow."

Emphasizing how much he has learned through the JGI trainings and through the app itself, Obed concludes, in a modest tone, "An innovation comes, and you are part of a team that starts with it, develops it. I think there's reason to be proud of it." With that, he and his colleagues return to their smartphones, applying cutting-edge technologies in the cloud to the realities on the ground, and to the daily labor of ensuring that humans and wildlife in the Kibale Conservation Area can continue to coexist.

Chapter 13

Outreach through fire

Dario Merlo hears Jane's words of hope as bombs fall on Goma

Looking out across the water from the eastern shoreline of Lake Tanganyika, the steely-blue forested skyline of the DRC promises all the relative tranquility of western Tanzania. With their respective shorelines separated by just 21.3 miles of water at their nearest point, it's easy to assume these two countries share many similarities. After all, this international body of water is home to the same diversity of cichlid fish; each water's edge harbors the common sandpiper and banded water cobra. Above, palm-nut vultures ride the transnational thermal updrafts, while below, chimps swagger and Nile monitors strut. But, beyond this overlapping ecology, the two nations have little in common. In the DRC, the context of daily life and the work needed to balance conservation and community development present unique challenges, challenges quite distinct from other JGI programs.

Dario Merlo was part of JGI-DRC from 2006 to 2019, implementing numerous programs in the war-torn nation, his country of birth. At the age of 25, he took on the role of socioeconomic development manager, and later became the institute's country director. He joined JGI in an uncertain time, amid civil unrest in the eastern DRC, a simmering stew of disputed transitional governments, guerrilla occupation, and the long-standing and contentious ethnopolitical Hutu-Tutsi conflict. "It's

Dario Merlo. *The Jane Goodall Institute, Adam Bean.*

complex," says Dario. "The realities in every country are different. There are basic needs we must respond to, and issues that need to be tackled. This can only happen when you listen to the people and understand the place itself. Then you can have an impact."

During a time when other conservation NGOs may have been reluctant to gamble their finite resources in the eastern DRC, the Jane Goodall Institute was steadfast in furthering its commitment to the region. From the outside, the DRC, a nation plagued by decades of civil turmoil and humanitarian struggle, might not have seemed like promising ground for conservation efforts. Yet in the shadow of war, JGI, led by Dario, began to effect positive change in the lives of a great many people, starting from the ground up, with the Roots & Shoots program.

"In some ways, I was learning as we were going," Dario admits, adding that he was fortunate to have mentors from various chapters of JGI who made the organization feel like family. "What truly impacted me and what I will always remember," he continues, "is the way we could change the daily reality of the people where we worked. We would arrive in some villages where the schoolchildren were sitting [outside] on the ground, under structures made of branches with banana leaves as a roof. Most of the forest schools, though, were made from clay-like mud, which the villagers had to keep fixing almost year-round because of the heavy rainfall. So really, there were no schools. We built them. Proper concrete schools, with dimensions and materials meeting the UNESCO [United Nations Educational, Scientific and Cultural Organization] standards," each with a playground and a small library. A mixture of empathy and pride appears on Dario's face as he reflects on these early days, all the more remarkable because, from 2005 to 2013, eastern DRC was at war.

"All the work that happened was carried out through a war situation," Dario says, "and our staff didn't do it for the money. When you do work like that, what price can you put on the safety of your staff? How much will you pay them in recognition of the great work that they all do? People do it because they truly love the organization, and they truly love their country. I would see [staff] go into the field, knowing that a day may come when one of us is kidnapped or even killed. But we managed to stay alive," he adds with a chuckle. "And by staying alive, we continued to have an impact in the remote villages we were working with."

Dario describes how the team would go into villages where people were drinking polluted water—so they built wells for those communities. "It was giving people dignity," he says. "Giving them their dignity. In a country where the people suffered so much because of what we all went through, the most amazing part was to give them the dignity of going to a proper school, to a proper health facility, to be able to access clean water. It's a good feeling." What kept his team going, Dario says, was a sense of ownership in JGI's holistic approach. "It's pride," he says simply. "It's part of the DNA, the culture of the organization. We loved our work, changing people's lives for the better."

As Dario reflects on that period, his mood turns pensive. "You know," he says, "the situations we went through during those years, I still think about it. I think about all the militia we would encounter, their barriers we had to cross, the complexity of reaching the field to continue our work…" A pause. Dario shakes his head slightly, as if still in disbelief. "You know," he adds, "we once used a plane that, only a year or so after, actually crashed because it just was so old." Again he shakes his head, this time with a smile. "Really, though, we had passion and that kept us going. We were going into villages and coming face to face with rebel militia. Our staff on the ground, they risked their lives to implement the program; they did it because they believed in the work and they loved their country—they were heroes. But when you get to teach youth, empower youth to understand that their lives can improve with an understanding of how to exist in their villages, with their forests, and why balance is so important, it's a good feeling. Because you know things can begin to change."

These young people will, after all, grow to become the community leaders of tomorrow. In Dario's experience, "everywhere the Roots & Shoots program was implemented, it was successful because it was an authentic entry point into a local community. It's authentic because you're not going to the local people to tell them what they need to do; you're really going there to talk. You come with subjects that are relevant to them," he says, "and for the environment. Do you tell them there is this many square kilometers of forest to be protected? No, you talk. You ask them why their forest is important. Is it something that they find important? Then, of course, they will say yes, because it provides shade, it provides them their medicines, and so on. There are always different things they will express to you, but it's authentic."

When relationships are built by engaging in open dialogue and understanding, a platform for inclusion can begin to take shape. "You can protect their forests with them," Dario explains, "because they now choose to participate, instead of using the typical NGO approach of telling them what the objectives are. When you protect their forests with them, you can talk about protecting the animals, which is the third major component of the program. In a soft way and without forcing it, you can explain why they should refrain from eating bushmeat.

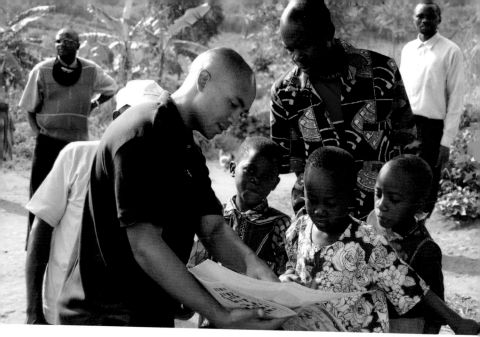

Dario Merlo (*left*) shows pictures of chimpanzees to children in Kasugho village, where JGI has built a micro-hydro power project. *The Jane Goodall Institute, Bill Johnston.*

"Of course, people like bushmeat, but we would start by discussing how every animal has a life, has a spirit, and you can see that when you are with your animals around you. You can see that even your chickens, they behave differently. If you explain more about the animals in the forest, what their roles are and the impact that they can have," it can change people's attitudes, Dario says, in subtle and unexpected ways.

As any teacher can attest, some seeds take a long time to sprout. But when these roots take hold, the shoots rapidly follow. "We had guys coming into the DRC from refugee camps in Tanzania and Burundi, our neighboring countries—then they created their own Roots & Shoots programs!" Dario exclaims, proudly noting that in eastern DRC, Roots & Shoots went from around 100 members to more than 25,000 in just seven years. "We were visiting every single school across the communities. As educators, we needed to teach them about the threats to the forests and the animals—why it's happening."

Despite Dario and his team's best efforts, an increasingly threatened DRC faces environmental pressures from an exponential growth in population and decades of on-and-off civil war. These both deepen the level of poverty and heighten the regional food shortage. Just beyond the sound of children laughing and singing in a Roots & Shoots class lies a backdrop of humanitarian distress. The ongoing crisis has had a cascading effect on the surrounding natural resources, specifically the forests and their wildlife.

The DRC holds the second-largest continuous rain forest in the world. To date, these habitats have suffered approximately 4.5 percent loss, an increasing and relatively recent trend. For a long time, these forests remained relatively intact, a stronghold for tropical Africa. But the growth in human population has been accompanied by increased harvesting of trees to meet the demands of impoverished people. Charcoal production, subsistence food crops, industrial agriculture, mining, and logging all checker the landscape with deforestation. Tree cutting for household firewood is also increasing rapidly and is one of the biggest causes of habitat loss. As village-adjacent vegetation recedes due to overharvesting, the topsoil dries out, and previously accessible surface water disappears.

Other habitat types such as woodlands, savanna, shrublands, aquatic grasslands, swamps, and extensive riparian zones are also under threat, as are the tributaries associated with the Congo River basin, due to pollutants. The DRC's eastern border captures about half of the Albertine Rift and a small portion of the Great Rift Valley. Given its overall size and equatorial location, its wide array of habitats, and unique geological formations, the DRC may well hold the richest biodiversity in Africa. Yet, as Dario describes JGI's conservation-science activities, the challenges of trying to carry out comprehensive, field-based activities in his country are all too apparent.

"Very few organizations in the past have done what we have done in the DRC on the conservation side. We brought science to places where nobody would go. There were forests, community forests, and national parks where nobody will go anymore because of the insecurity. For most of that forest, few studies have been conducted in recent years. We came up with the very ambitious program of putting together a DRC conservation action plan (CAP). Our aim was to combine all the different conservation actors and stakeholders to define a common strategy, a common goal. Some of the challenges were to find the number of the chimps via a census and identify their threats. We used the Open Standards for the Practice of Conservation as part of the JGI Tacare approach and ended up with about 20 different partners from government, other NGOs, and from the private sector.

"Putting together a CAP in a country that was devastated by war, when all the other NGOs were working on their own, not really sharing information, not really reporting to each other or sharing data—man, that was a real success," Dario reflects. "I think at the time JGI were the only ones with the capacity to achieve this, because we were an organization that had a great culture of sharing and not withholding anything from others. We wanted it to be low profile, but effective."

In the DRC, the areas for community intervention were determined by JGI's conservation targets, specifically great apes and their habitat. The teams based their research work and scientific data on the fieldwork of their partners, because, at that time, JGI did not have the capacity to do biomonitoring in the DRC. After the CAP was completed, JGI became more active in the area, in partnership with the Congolese Institute for Nature Conservation (ICCN), a government organization, and UGA-DEC, a nonprofit dedicated to gorilla conservation and community development. Together, Dario says, they formed a good team, and their actions had a significant impact on the protection of chimpanzees and their habitat. Even today, with the 2011 CAP, updated in 2015, as a reference for other conservation work, Dario believes that the DRC has the greatest potential for protecting this endangered species.

The country is home to approximately 40 percent of the global chimpanzee population, and the DRC's northeastern forests represent an important stronghold for the species, along with their great ape cousins,

Participatory mapping of expert knowledge is part of the great ape CAP process in eastern DRC. *The Jane Goodall Institute, Lilian Pintea.*

the Eastern gorilla. Yet, despite its biological significance, the eastern DRC remains under a cloud of virtually constant risk to human life. "Let's say you applied [standard] USA risk management processes to another country, like here in eastern Congo," Dario says. "All of the work we have achieved, none of it ever would have happened, because the risk management process would never have allowed for it. It happened because we took risks, and we understood the complexity of the context, and we dealt with it.

"That's what makes JGI's Tacare approach unique," he continues. "If we had brought in people from America, or Australia, or Namibia and put them in DRC, and said, 'Go out and visit schools in an active war zone,' nobody would go out and do the work!" JGI's holistic approach to community conservation is effective, Dario believes, "because you have people from the country taking care of it, and they do it with their hearts. Of course, you can talk about good management, setting your goals, your objectives, measuring, etc.—you can talk about all of that, but [the key is] that JGI has a lot of staff members working for their home countries, and so results follow."

The Tacare approach depends, to a great extent, on having the right people in the right place. As Dario points out, "You can have the right people in the wrong place, and progress will stop. When you look at all the people Jane has been with for years across her projects, in Tanzania, people like Mtiti, Dr. Shadrack, Dr. Anthony, Lilian, and Mary Mavanza, and then in Uganda people like Peter Apell, and there are many more—these are the people that have made JGI work and are there because they want to change their own country for the better."

Here, Dario pauses for a moment, with a grin. "And she does it too! Jane, I mean. She puts herself at risk because she truly cares about her people. I had tried to have Jane come to DRC for many years beforehand. I wanted this because I thought when you have the founder coming into the project country, it's just a superb recognition for the people who are doing their jobs. We knew Jane would go to Tanzania for obvious reasons, to Uganda, sometimes in Congo Brazzaville, but not the DRC. I thought it was because of the insecurity—it was not the safest thing for her to do. But finally, in 2012, Jane was coming."

While Dario was in the process of arranging for Jane to visit the DRC branch, an undercurrent of the DRC's political unrest was resurfacing. Later referred to as the Kivu conflict, this decades-long crisis consists of two broad influences that are intertwined, each perpetuating the other. The first is the economically driven conflict mineral trade, and the second is the political and civil unrest, much of which follows from the 1994 Rwandan genocide. The ongoing trade in conflict minerals mainly involves coltan, an essential mineral for the manufacture of technologies such as mobile phones and computers. The DRC's Kivu coltan deposits are the largest in the world, and with an extensive Western market, this trade both finances and complicates the second factor, war.

In 2012, the leader of a breakaway group, the National Congress for the Defense of the People (CNDP), mutinied against the government,

DRC villagers in 2004, during a site selection trip. *The Jane Goodall Institute, George Strunden.*

claiming that a 2009 peace treaty had not been honored. The CNDP's followers formed a militant group, the March 23 Movement, or M23. While Jane was on her way to Goma in the DRC, M23 militia were positioning themselves to capture the city.

"We couldn't obtain a visa for Jane to come in," says Dario. "The political situation was getting really bad, and the Congo Embassy were not issuing visas for visitation. The embassies, especially the British Embassy, were telling us that Jane should not travel to the DRC, especially to eastern Congo because there were conflicts breaking out. Artillery fire was happening just outside of town, about 5 to 10 kilometers away. There was a rebel army, and they were coming into Goma. Jane still wanted to come. I was worried, and I kept wondering if we should continue with the plan for her to come."

Yet, says Dario, "she insisted. She said, 'If you live there, then surely I can survive a visit for a few days. If you are living through it with your wife and child, and working through it, why should I be forbidden to come?' I knew we were taking a risk. But I also knew that if she came during that time, it would be sending a strong signal to all the organization. Jane is a superhero. She had no fear. The management people almost never came to DRC because they feared the political and security situation. But we got the visa, and Jane came."

During Jane's visit, Dario arranged a meeting with one of the biggest Roots & Shoots groups at a local university. By that time, he says, "the fighting between the government and the rebel militia was taking place around the outskirts of the city. We could hear the bombs dropping. The soldiers came and Jane kept talking, inspiring people. There were a thousand kids in that stadium, then we had another group listen to Jane in a basketball stadium.

"Many kids who were in these stadiums," he adds, "about 20 of them, they are now in the US; they're studying out there. I remember she kept saying, 'We are going through some really tough times, but we should have hope for the future, and we should work for a better future. How can you change the world we live in?' She was saying that while we could hear the fighting all around us. At some point, we had to evacuate because the bombs were coming too close to where we were. We had to leave." Jane was evacuated to a somewhat safer location but went on to spend two days in Goma. "It was complete insecurity," Dario admits. "Anyway, Jane left and then a week later, the rebel militia took over the city. They just took it. It was fighting unlike anything I'd ever seen before."

Dario says that Jane kept calling him during that time and sending messages, to find out if he and his family were okay. "But with the rebel army fighting with our government in town, I was…" Here, Dario pauses, as if perhaps reliving more than he wants to. "I couldn't leave because all the roads were blocked, the fighting took over. Of course, it was tough, I really thought I was going to die at that time." Again, he stalls, then calmly adds, "I've been through two different wars. In 1994, the Rwandan genocide affected the region because civil war overflowed into the eastern DRC. Look," he says, returning to the topic of Jane, "what makes

A soldier in war-torn eastern DRC in 2004. *The Jane Goodall Institute, George Strunden.*

our organization unique is that we have someone who cares. Jane cares for people. For her, it's not just about how good you are at implementing your program. It's also about how good you are as a person, and she cares about you, your children, your family. I don't know that many founders who really care. You can see that she cares for you as a person. For example, even now that I am no longer a JGI staff, she still cares. Our spirit goes beyond the professional relationship. That's what makes the culture of JGI. It's part of the whole organization."

Jane's philosophy of holistic, locally driven change is one that goes far beyond a flashy logo and a mission statement—because Jane leads by example. The JGI staff on the ground echo these lessons by fostering an approach that relies on the genuine engagement of local people, and a willingness to hear and learn about their realities. By aligning the humans' struggles with the struggles of the surrounding plants and animals, the Tacare approach and programs such as Roots & Shoots can be adapted and applied in the context of the eastern DRC. Even in a war zone. Even through fire.

Chapter 14

The banks and the bees

Phoebe Samwel links microcredit to women's empowerment

KICODA harvests honey—and venom

As the story of Tacare has shown so far, the balancing act of social development and sustainable conservation is a delicate one, and success for one may cause hardship for the other. Yet, for a snapshot of the viability of both ecological health and human community development, perhaps we need look no further than a community bank and a beehive. Consider their similarities: A community bank reflects the socioeconomic health of a developing region through the total assets held, the number of new accounts opened, and the volume of activity in and out of the bank's front door. Likewise, a beehive holds the sweet wealth earned by each worker's efforts; the volume of honey hints at the health of the surrounding vegetation; and the number of workers passing through the front entrance shows whether the colony is active and thriving.

Since 2015, JGI's Phoebe Samwel has been closely monitoring the steady rate of success from microcredit initiatives supporting small businesses and alternative incomes in Kigoma, Tanzania. As a community development officer, Phoebe provides small-scale entrepreneur training to village members trying to break the cycle of resource-damaging subsistence living, while also providing guidance to those wishing to save their money using the COCOBA.

"My role is to work with people and help them promote livelihood activities," Phoebe says, "with a goal of changing their livelihood status" to give the environment a chance to recover. "Unless income security or food security is accessible to the people, conservation initiatives can't advance under the continued pressure of resource overuse by people. My goal is to see people earn money by increasing their environmentally conscious income-generating activities, which then increases the likelihood of their children receiving an education. That is my big role."

Phoebe reiterates Tacare's basic principle that trying to address issues such as conservation or education with local people is pointless without first addressing the issue of livelihood. "In this landscape, many people are surrounded by natural resources. It's easy for them to cut trees to sell for money, and for so many people, it's all they can do. But for the conservation of biodiversity, and the health of their environment, this is not working. We need to train the people in generating alternative incomes that are environmentally responsible, so that each side endures and importantly, also improves. My role is fitting this in everywhere. Even if you promote family planning, you also need something to eat. Even if you want to conserve trees, you still need to use wood for cooking. My work is cross-cutting because every human being needs money to survive."

In advising and training those who wish to generate income or start a business, Phoebe says, "We show them how to manage the small funds they have through different activities to promote future income. They need skills to manage money and more importantly, they need skills to manage credit. With the loans they get from microcredit groups, we teach them how they work and what is needed to start a business, repay the loan, and generate a profit for life."

The Tacare project began by promoting savings and credit cooperatives, a microcredit program managed by the district cooperative officers. The project was piloted in 10 villages and did well, but after a while

Phoebe Samwel, JGI Tacare community development officer, stands in front of a group of microcredit beneficiaries in Kigoma, Tanzania. *The Jane Goodall Institute, Shawn Sweeney.*

Tacare needed to supplement the funds for members to borrow when starting their businesses. So COCOBA was introduced, with around 30 members at first. Soon, more people joined and, with training, started to save. "It's monitored by the group members themselves," Phoebe says, "which means it's not bureaucratic. It's a friendly methodology in which they can start and end [memberships] after 52 weeks, open or close memberships easily, and receive loans without any complications. They own it themselves."

Once the locals joined, Phoebe explains, they were trained in entrepreneurial skills, which many of them used to start income-generating activities, such as opening small pharmacies, food kiosks, or shops to sell necessities like sugar, salt, and soap in the villages. "After our first month into the project, we went back and found progress," she reports. "People are changing and they're making profit, some are buying land and others are improving their houses, but in general, people are happier. Which touched my heart to see the community happy, incredibly happy. When we ask why, they say, 'Now we can get our money without any problems, any time we like. We can run our business. We can do loan management ourselves.' To me, it seems to be bringing about big changes in this community, especially for women, even though the COCOBA group membership is still around 80 percent men."

For women, and entire households, Phoebe notes, the changes in livelihood have meant a change in status. "Women who were voiceless now stand and speak in front of the village assembly meeting. They can contest for the leadership position in the village. They can own their own land. The sense of ownership now is increasing, and the sense of trust is increasing. The sense of participation in decision-making at the village level, at the ward level, is now very high also." This increased empowerment of women means a corresponding increase in youth empowerment, she says, "because many youths are accompanying the women to the COCOBA and start saving their money also. They now know to plan a simple workplan and a simple business plan, and they can do it themselves. For us, this is a big change."

As a result, according to Phoebe, fewer trees are being cut down, less charcoal is being burned because COCOBA members are using fuel-efficient stoves, and many people who once relied on poaching no longer need to. "One of the laws to be a COCOBA member is that you must agree and sign that you won't use money from COCOBA to pay for any activity that's not environmentally friendly, like selling charcoal, like cutting trees. You can't get a loan for doing those activities. Also, more people are becoming involved in beekeeping activities because they get an income from their hives." The COCOBA project supplies beehives, supports training for the beekeepers, and helps them find a market for their products.

Phoebe cites many examples of women who started a business and bought land, constructed their own houses, and sent their children to school. "I also have a list of men who are now changing their behavior," she adds. "They've also started businesses and they've started to help their wives and send their children to school. That's a noticeably big shift in social behavior, especially for women. Families have started to send girls to school, whereas for many years it was not a priority to send the girls to school; the priority was the boys. Nowadays, you can see more girls getting an opportunity to receive an education."

When Phoebe wondered how this was possible, she learned that "in 1998, Dr. Jane brought a friend called Deborah Simons here, who started to sponsor girls' education. Nowadays, we have more than 700 girls who have been supported by this scholarship and completed their higher education. We have doctors; we have nurses; we have teachers; we have many wildlife people," and many of them are coming back to their villages to encourage the communities to support education for girls.

For Phoebe, what's important is that programs like COCOBA or microcredit are self-sustaining. "The sense of sustainability is there because these people are achieving their own results. The result is not ours; the result is theirs. They own the project activities because the project was developed with them, not by us. We work for them; they are not working for us. Even if we can't visit some of these communities enough, they continue working, and we hear these stories. That is the sustainability that we like to see, because we may not be here forever, so we must make sure they're capable. And train them so that they can keep this capacity even over time."

Avoiding what Phoebe calls the "syndrome of dependence" is crucial, because otherwise, she says, you'll invest a lot of money and end with negative results. "But if you make them own their result, own their project, their approach, and then withdraw just enough to monitor things, you'll see. That's why we train the forest monitors. We train teachers of teachers (TOTs). We train them on COCOBA because they take our roles, our responsibility. They are becoming their own teachers and becoming the trainers now." Also, Phoebe says, technology such as smartphones is

A shop owner beneficiary of a JGI Tacare microcredit loan. *The Jane Goodall Institute, Shawn Sweeney.*

enabling villagers to record the trainings and train one another. "I can see communities are now using this technology to spread news and to share. How they share the information like this is a big advance on years ago, and I want to see sustainable communities continuing to work with us and share all our investments with others. After all our time we invested, I want to see how the community goes forward by itself. I want to see women speak, [and] women empowered economically, politically, financially. I want to see women change in the community, because if women change, the family changes also. Girls change, boys change."

Clearly, this is a topic close to Phoebe's heart, and she continues, "I don't want to see voiceless women again. I want to see women who are talking on behalf of their children, on behalf of their husband, on behalf of their community. I don't want to see these [previous] issues coming back. Now, I'm seeing it with my own eyes. I'm seeing the community sustainable, community planning for themselves. I want to see more communities like that, who are owning the process."

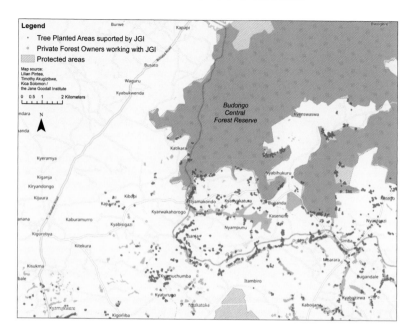

The map shows Budongo Central Forest Reserve in Uganda, including surrounding local communities, PFOs, and tree-planting efforts by JGI using the Tacare approach. *The Jane Goodall Institute, Lilian Pintea, Timothy Akugizibwe, and Kica Solomon.*

When Phoebe speaks of the benefits arising from microcredit and banking opportunities, her sense of pride, particularly in the empowerment of women, is infectious and heartwarming. Yet, in true Tacare fashion, these social adjustments go on to benefit the entire household, relieve the financial burdens of a single income, and reduce the extensive natural resource use and degradation in these areas, allowing time for the ecosystem to respond and recover.

Just over the northern border of Tanzania, in the Hoima district of western Uganda, another group of remote villagers has also successfully adopted sustainable livelihoods as entrepreneurs, community forest monitors, and beekeepers. Their progress is such that, today, a byproduct of their farming is even contributing to medical research.

Members of the Kapeeka Integrated Community Development Association (KICODA) group. *The Jane Goodall Institute, Adam Bean.*

A villager named Mustafa introduces himself as the general secretary for the Kapeeka Integrated Community Development Association (KICODA). "At the same time," he adds, "I'm also a poultry man and a bee farmer." KICODA was launched in 2006 to develop environmentally sustainable income alternatives to reduce the deforestation of village lands, while also providing local support for the Ugandan authorities' management and conservation of the Budongo Forest.

Mustafa, who is studious in manner and deliberate in his choice of words, tells the story from the beginning. Around 1995, he says, the National Forestry Authority (NFA) proposed a plan to form Collaborative Forest Management (CFM) groups and to consult with the communities on the sustainable use and management of forested areas on community land. The community then developed an agreement with NFA on methods of managing these resources for environmental and economic benefit.

John William Mukongo showing some of the beehives he received from JGI Tacare. *The Jane Goodall Institute-Uganda.*

"We have a binding document that guides our activities in line with the constitution of Uganda," Mustafa explains, "and our CFM is registered and has become a model CFM in Uganda because of the activities we're doing in the forest." Before this intervention, the forest was "in a critical condition," Mustafa says. "But after the NFA approached us, through sensitization and education activities, communities started changing their attitude towards conservation, realizing that income opportunities could be developed without the need to further degrade forest resources. The NFA permitted us to carry out projects alongside the park boundary, so we conducted bundle planting along the forest, where we plant our own trees for harvest. On individuals' land, we started sensitizing communities to also do other private forest-related activities like goat rearing to improve their livelihoods. One of the core objectives was to improve the livelihoods of local communities at a household level."

At first, Mustafa says, his group had five members, but when JGI came in, another five were motivated to join. This group teamed up with

NFA to handle illegal activities in the forest, and, in return, NFA provided them with 400 beehives, as an incentive to generate alternative sources of income. In consultation with the villagers, 40 community members were selected to become bee farmers and provided with 10 beehives each.

Mustafa describes all the other equipment available to the bee farmers: "We have the honey settling tank for storage of honey after harvesting and processing, which can be stored for up to two years. We have the honey press machine; this is used for squeezing honeycombs to extract the honey. We also have the candle molding machine, which produces these candles from the wax byproduct of processing. Other equipment like the honey refractometer is used for detecting moisture content in the honey, which allows us to keep the quality high when combining honey from other bee farmers, ensuring that the moisture level isn't too high, because that lowers the quality. We also have the venom-extracting machine, which is used to collect bee venom, as it has scientific and research value."

Essentially, a venom-extracting machine is a mounted piece of glass the size of a small picture frame, with rows of electric wires strung inside the frame. As the bee lands on these wires, it is stimulated by a small electric current and responds with a defensive sting to the surface of the glass. Though the stinger does not dislodge from the bee, it releases the venom, which is left on the glass. The release of alarm pheromones by a bee "attacking" in response to the electric current calls other workers, who rally and join the defense. Soon, the glass is laden with their venomous residue, which, once dry, is scraped off and collected in a crystallized form for easy transport to a medical research facility.

Bee venom is used in many medical applications, including the treatment of HIV. Scientists have discovered that a compound in bee venom called melittin passes through the compromised membrane of HIV-infected T cells, destroying the cell from within. As healthy T cells still have their outer defensive membrane intact, the melittin from the bee venom cannot penetrate them, meaning that bee venom kills only the T cells infected with HIV, leaving the heathy cells unharmed. This

Mustafa shows the venom-extracting machine designed to safely collect samples for medical research. *The Jane Goodall Institute, Adam Bean.*

unexpected byproduct of beekeeping shows how community initiatives like KICODA can benefit programs or projects that might otherwise seem completely unrelated. Such synergies also support the cross-disciplinary approach known as One Health, working locally, nationally, and globally to attain optimal health for people, animals, and the environment.

When the beekeeping business began, Mustafa says, members of his community were given training not only in beekeeping but also in marketing, leadership, lobbying, and advocacy. With these skills, they've gone on to train others, and even non-beekeepers are benefiting: "They are beginning to understand that it's not just about getting honey, but that these bees are pollinating crops, making people more appreciative of their value. Everybody is much more aware here that the bees are the world's number-one pollinators." Meanwhile, the community of beekeepers benefits from the transfer of skills, education, and equipment.

"It's called farmer-to-farmer extension services," Mustafa explains, "and part of that is helping them identify whether their hive is ready for harvesting. We have a special team for that, to ensure the quality of honey and the harvesting process. When it comes time for harvesting, our cooperative here buys their honey, which we sell in town here in Hoima, and we also have buyers in Kampala."

However, like all farming, beekeeping is subject to seasonal ups and downs. "We harvest in March and in September," says Mustafa. "Our last early season when we collected in March was productive, and we collected around 200 kilograms. The challenge we faced was in the second season, after we had very heavy rains. We really couldn't collect anything, so had to concentrate on what we had saved in storage to fill our orders, which eventually all ran empty. Those who used to supply us from other areas also had the same complaint."

Though their business is frozen for the moment, the beekeepers are hoping for more from their next harvest. In an optimistic tone, Mustafa reports, "The colonization rate has really increased during this last dry season and it's now very high. Of my 15 personal hives, only 6 were colonized after the last season, but now I have 13 that are occupied. This is promising and we are really hoping to see more of that, because we need this next harvest to be productive. Because of how the climate changes, and because bees aren't always predictable, we are working on other ways of generating business, so we aren't so reliant on just honey."

Such projects might include collective participation in buying and selling surplus products such as maize, Mustafa says. Also, with projects like tree planting, if everybody in the community has a plot, they can buy and sell the wood from mature trees to pay for expenses like their children's school fees. "The community needs to be able to sustain these forms of business, so our livelihoods at the household level improve. We all have success stories now which we didn't have before finding alternative sources of income," he concludes with pride.

One such person with a success story is Fred Lemerega, the chairperson of KICODA. "I joined the beekeepers in 2012," Fred explains, "and I also became a forest patrolman in the CFM group. I've seen the past, and I see how things are working now." What works, Fred says, is the community's collaboration with organizations like Budongo Conservation Field Station (BCFS), based on a shared interest in protecting the forest—which also entails offering community members sources of income other than cutting down trees or poaching wildlife.

"When the BCFS came in, they supported some of our farmers by giving them goats, especially those along the forest boundary," he says. "These farmers would complain about the impact from the animals, and the human–wildlife conflicts that occurred whereby animals would come and raid people's crops all the time, so people would take these matters into their own hands and kill any animal destroying the crops. In response, BCFS gave their support in the form of goats, and also started giving out seeds for certain crops that wouldn't be eaten by these animals, such as soybean, cabbages, and other vegetables. Community members started realizing that when we worked hand in hand with those partners, things could improve."

Fred speaks from the heart when he describes the transformation in his own life that KICODA has catalyzed. "For me personally, before I joined, I actually had nothing. I was somebody who would go through the day without any timetable. When I joined, I realized that I could use my time to maximize my opportunities and take control of my circumstances. This is only possible when our partners helped us implement alternatives, because before that, we had no other way. Last season I harvested 50 kilograms of honey, which I sold and managed to pay the school fees of my daughter," with some funds left over for new clothes.

Yet, before KICODA offered alternatives, not all villagers were surviving by subsistence farming. One of the livelihoods most affected when community conservation measures are put into practice is that of a career poacher. Although outsiders with limited understanding of the day-to-day realities of these remote villagers might be quick to judge, when a person has limited resources and an adjacent protected park full of bushmeat, poaching may be the only way to survive. For this reason, many global sustainable initiatives are aimed at those who rely on poaching, not only to reduce the impact on ecosystems but also to remove the potential threat of disease transmission between people and animals, otherwise known as zoonotic diseases.

Moses, a former poacher, is not a member of a wealthy international syndicate for the trafficking of animals or animal products; he is a rural man who has, in the past, relied on poaching to feed his family or to pay for necessities in the village market. With new livelihood opportunities, Moses has happily made the transition to a better way of life—one that doesn't carry a sentence of life in prison.

"The poaching work," says Moses in Swahili, "had a lot of challenges because we might be in the forest for many days at a time, and it would take a lot of planning. I would need to carry up to five days of food and would have no idea what animals would be there, which was extremely dangerous." Although understandably reluctant to give too much detail on his time as a poacher, Moses' age, strong frame, and weather-forged features suggest a wealth of knowledge about forest ecology.

"This forest has many animals," Moses continues, "so you may be looking for antelope when you find lions instead. We were risking our life every time we went in. When I heard about KICODA, I decided to stop, because my livelihood then was all about the risks to my life in the forest, and then the risk of being arrested and placed in prison. If the rangers found poachers in the parks and they tried to run, they would be gunned down. I looked at my family and said, 'Now I can do away with that life.' So now I am free from those challenges, and I have an incredibly happy life. I grow some trees in a small plot, grow my cabbage, and I have some beehives. I have even been able to buy sheets of iron for a better house. There are now monkeys that come around to my home to eat my pawpaw, and I stop the children from harassing them. My life is much better now."

If someone told you that a single community development group was able to protect the forest, create plots for sustainable timber harvest, generate alternative income, assist the national government with anti-poaching patrols, sell locally harvested products, and help the broader One Health approach with HIV treatments, all simply by bee farming, you'd be forgiven for thinking that was impossible. But the fact that it *is* possible demonstrates the connectedness of everything—people, plants, animals, and disease—and the power of holistic community-led approaches such as Tacare.

A bank and a beehive. Each a reflection of hard work, each acting as a central storehouse for the wealth generated by the community's commitment, each a sign of sustainable efforts to manage the resource competition between people and their environment. Like beekeeping, community microcredit initiatives like COCOBA have been groundbreaking for poor and isolated communities in Africa. Because these banks offer accessible financial management, specific to the needs of rural African communities, people's relationship with money begins to change. Poverty isn't just a cycle of struggle and daily hardship: it is a trap of only living in the here and now, an existence where securing the basics for survival becomes all-consuming. Without financial security, each day is taken as it comes, making the notion of planning for tomorrow, or next week, or next month almost inconceivable. Microcredit programs offer a chance to break the cycle of necessity and allow for choice; even if the daily income of a household doesn't increase significantly, a microcredit loan for a small business generates opportunity and most importantly, hope—and hope is a place where tomorrow, next week, or next month does indeed exist.

Chapter 15

Changing the retirement plan

Sania Lumelezi manages difficult conversations about choice

E arly in the original TACARE project, family planning was identified as a crucial area for community development. This was in recognition of the high birth rate and average family size in rural African villages, and the hardships large families impose on poor, resource-deficient communities, increasing the ecosystem degradation at the human–wildlife interface of these ecologically significant and biodiverse areas. Participating with Tanzanian health officials, JGI began women's health and family-planning outreach in 1999, sending Community-Based Distribution (CBD) staff into villages already familiar with the Tacare project, to begin sensitization, conduct household surveys, and develop a baseline response from the local women as to whether or not they would be open to discussing contraception and family-planning principles.

The deeply personal subjects of sex, marriage, and reproduction are, of course, fraught with potential for miscommunication and cross-cultural misunderstanding. It's no surprise then that the topic is a precarious tightrope to be negotiated when NGOs engage with local villagers on the subject of reproductive health, contraception, and the long-term benefits

Sania Lumelezi (Mama Sania).
The Jane Goodall Institute.

of having fewer children. Depending on their cultural backgrounds, some women might not understand why others choose to have seven or eight children, while others, in turn, might find it incomprehensible that a woman would wait until the very end of her reproductive life to conceive her first. What's important, however, is that childbearing is a choice and that the right to choose is a focus of family planning.

As this is a potentially inflammatory topic, JGI CBD volunteers are all local and work in their own villages, where they are known. That said, it is not lost on the people of western Tanzania that most contraceptives are a product of Western science, leading some to believe they are racially motivated tools designed by "white people" to "end the African race." Whether contraceptives are wanted, well intended, or refused, engaging people one on one is the only way to find the answers and maintain the right to choose. Committed to this approach is Sania Lumelezi—Mama Sania, as she's locally known—the family-planning coordinator based in Kigoma, western Tanzania. In this role, Mama Sania carries herself like a natural caregiver, a matriarch in her community.

Mama Sania was one of the first to start working in family-planning outreach for JGI, back when the Tacare project only operated in a few villages. "When we first visited," Mama Sania says, "we sat with them and introduced family planning to the families, talking about the benefits, and why they should be aware of what family planning is, and what the options were. We started with these villages first because they were

JGI Tacare health education and family-planning program in Kalinzi village. *The Jane Goodall Institute, Shawn Sweeney.*

surrounding Gombe. Secondly, they used a huge amount of the greater Gombe landscape for farming, cutting down trees for firewood and for building materials. As the population grew larger, the amount of land needed increased extensively. Here in Kigoma, they have shifting culti-vation, meaning they farm here this year, next year they move to another area for farming, leaving this one to regenerate the nutrients again. By doing so, they're also destroying more and more of the area."

As Mama Sania explains, in Kigoma most women have an average of six to eight children; having a large family is "a sign of prestige." But, she says, "most of the people are extremely poor. They can't afford to have such large families; they can't afford to give their children good food, an education, or keep them healthy by attending the dispensary for medicine when they're sick. Most of the children are not going to school. They're just at home because their father cannot afford for all of them to attend, and the ones that are sent are usually the boys. Girls are the last to be cho-sen for school," but, for financial reasons, that rarely happens.

"Also," she continues, "the marriage rates are high for these girls, and usually by 14 to 16, they're married. Soon afterwards, these young married girls start having their own children, so the numbers are always increasing. Again, we explain what a family plan is, and we highlight the benefits and recommend they start. We explain to them that the women should increase the time between children to regain their health after childbirth—normally they birth almost each year after their first child. We often see families with children aged 1, 3, 4, and on, from the same mother. The spacing between one and the next is very limited. We try to tell them, 'If you start family planning, you will also regain your health. After regaining your health, you can also help your husband. You can earn money to help your family.' As this goes on, we also try to explain about the Village Community Bank (VICOBA), a microcredit or lending program for business assistance."

According to Mama Sania, mothers who use family planning have time to join the VICOBA, which helps provide for their children. The father also benefits, because if his children are healthy and his wife is healthy, he has more time for work and needs to spend less money on medicine and food. Yet broaching such topics often leads to difficult conversations. "When I first started in 2001, there were arguments," Mama Sania admits, "the majority from the Roman Catholics and the Muslims in the village, because, due to their religions, they did not like family planning. But now as time goes on, people understand, and the program has even expanded."

From an initial 16 or 17 villages, the program has expanded to more than 60 villages throughout western Tanzania. Though Mama Sania is the coordinator in each village, the program depends on volunteers, who are selected by the village leaders. These volunteers are then trained according to a national Tanzanian curriculum for family planning; once trained, their responsibilities include data collection at the village level. "For each village, we are collecting data for how many people are using a plan, how many new users or old ones, and how many are returning to a plan. The volunteer will know who in their village is using family planning, and these numbers are reported monthly. At the end of the year, we gather all the data together and see what villages are increasing in their use of family planning and find out what is changing—if it is numbers of new people or if we need to do more outreach and sensitization in that village."

To date, Mama Sania says, most of these villages have been informed about family planning. On top of that, her teams are also educating these communities about conservation and protecting the environment: "To reduce the overharvesting or deforestation of trees around village lands, we also provide training for how they can build their house out of bricks and explain why they should keep a small shamba [plantation plot] at their home for growing their own trees for future firewood use or as a sale crop."

By now, it's clear to Mama Sania that people are beginning to understand the long-term benefits of family planning and sustainability. "Those who were among the first to start family planning and living with sustainable private shambas are now showing the rewards," she says. "The trees are grown up; some harvest trees and give them to their children. Others they're selling for a lot of money to send their children to university. These earliest participants now have money, and they're building modern homes in the village. The women now are healthier. They can have a shamba; they can have trees; they can join VICOBA. The people who were rigid and against family planning, they can see their neighbors; they can see their health and how their children have been sent to university. They see all these benefits and now, many [people] are coming. We now have a shortage of certain contraceptives because so many women are seeing the benefits and coming to say, 'These children I have are enough for me,' or men saying, 'I don't need any more children and I want my wife to relax,' and getting a vasectomy. We are giving these people back their choices."

Mama Sania credits the teamwork between village leaders and CBDs in some of the villages with helping to keep the program on track. "It's like how Tacare is supposed to work," she notes. "It's a team effort, and now the leaders understand. When I go to the villages to meet with the CBDs, I report to the village leaders, and he or she can inform me of how the CBDs are working, or make suggestions for a new CBD, or inform me if a CBD is not working well. We also show them the reports we send to the project, so now they're aware. For example, we had one CBD who went to a pastor's wife to give health education about family planning, and the pastor became rigid and was very angry and rushed the CBD, chasing him with a knife. Our CBD made a report to the village leader, so we know they're being protected."

Among the challenges that the program faces, Mama Sania says, is finding ways to keep the CBDs motivated. Another is the method of data collecting. The family-planning data is still collected manually, but Mama Sania would like to introduce technology like the smartphones used by the forest monitors. "These would serve as good motivators for the CBDs and make data collection more efficient," she explains. "Then we will have data and reporting in the main computer system." Though up-to-date technology would enhance the teams' effectiveness, on the whole, Mama Sania says, "I think we are doing well with what we have."

As Mama Sania's experience shows, family planning is still a sensitive topic, even in communities that have been participating for years, and cultural change will be a gradual process. But in terms of environmental impact and quality of life for families, it is one of the most significant and potentially impactful projects under the Tacare umbrella. The global population increase is exponential, putting devastating pressure on economies and natural resources. As with many other social development programs, meaningful change and tangible results are not imminent and will likely come down to a gradual shift over generations.

Arguably, one of the most common rationales for such a high number of children per household in rural African nations is the need to secure help with labor and care for elderly parents. In wealthier parts of the world, aged care is something that is provided or at least available. Health care services in such places are also advanced, providing age-appropriate care and treatment. Poor African villages do not have retirement homes, nor do they have anything that resembles age-appropriate medical facilities. In addition, most people in these areas who reach an advanced age have worked a physically demanding life, and it's not unusual to see elderly and visibly disabled people carrying water or plowing a field by hand. Suddenly, having a small team of younger adults to take care of their parents may not seem like a limiting decision after all. It may be that the key lies in reducing large families but still having enough children to share the responsibility of caring for their elders.

For now, these families still face a difficult choice. Do they invest in large families, with the added benefit of having multiple children and grandchildren around to help with household needs and provide care and support when the time comes? Or do they provide a higher quality of care to fewer children and put their surplus resources toward building a higher standard of living and perhaps greater future security? As always with Tacare's initiatives, if Mama Sania and her CBDs continue to engage, listen, understand, and facilitate, the answers will emerge from the communities themselves.

Chapter 16

Of women champions

Alice Macharia paves the way for African women in conservation

Yakaka Saweya explains why so many village girls don't complete their education

When seen via satellite imagery, Lake Tanganyika appears as a thin rip in the fabric of the African continent's landmass. Humbler in appearance than Lake Victoria and less visited by tourists than Lake Malawi, Lake Tanganyika (shared by Tanzania, the DRC, Burundi, and Zambia) promises little more than a scenic strip of water captured in the creases of the Albertine Rift, a drop in the cupped palm of Africa's hand.

Nevertheless, under the surface unfolds one of the most significant narratives of biological speciation that nature has to offer: the diversification of its cichlid fish. With more than 200 cichlid species described in the lake and an additional 50 or still undescribed, these fish represent how adaptive radiation of a common ancestor, or ancestors, can lead to variations in niche selection, morphology, and physiology, and ultimately, to species divergence. Although experts are still debating just how many cichlid species Lake Tanganyika contains, and whether their variations are mutations, location types, or definitive species, what's certain is that their differences have them occupying all depths of water, feeding on all

Alice Macharia (*left*) and Yakaka Saweya, deputy head teacher and senior teacher of female students (women's patron) at Wambabya Primary School. *The Jane Goodall Institute, Adam Bean.*

possible aquatic diets, and presenting distinctive sizes, colors, patterns, and behaviors.

Perhaps more impressive is that 99 percent of Lake Tanganyika's cichlids are endemic and not found anywhere else in the world, not even in neighboring East African lakes. This phenomenon may be because Tanganyika is the oldest of Africa's lakes, or because it's the deepest, or because it has the largest volume of water, or because it has a more suitable array of microhabitats or some other unknown cichlid-friendly variable. Whatever the reason, these fish illustrate how diversity equals entirely different ways of living, even within the confines of a single place.

On a much larger scale, the diversity of humans across the globe also translates into different ways of life—between regions, villages, and even genders. Alice Macharia, vice president of Africa Programs for JGI, who oversees the institute's activities across Tanzania, Uganda, Congo, and the DRC, has seen for herself what such geographical and cultural variations mean for JGI's projects in program countries.

"Previously, I'd done research in Kenya," Alice says. "I was looking at wetlands and comparing two sites, Lake Naivasha and the Tana River Delta in Kenya. One element was a comparative study, looking at how

A school of juvenile giant (emperor) cichlids (*Boulengerochromis microlepis*). A contender for the title of world's largest cichlid, these are endemic to Lake Tanganyika and so not found anywhere else in the world. *The Jane Goodall Institute, Bill Wallauer.*

or if the [local] people were involved in conservation. At one site, people were so involved, they were even directing research questions that researchers could pursue. With local knowledge, they would notice things in the environment, and this would guide research in that direction." At the other site, though, the community's level of involvement was less impressive.

"In 2003," Alice continues, "I was doing a master's at Brandeis University. For my research paper, I was very interested in how gender had been integrated into conservation programming, or which institutions were working on conservation but with a strong community focus. I'd been talking to a number of people and my aunt said, 'There's this organization called the Jane Goodall Institute, you should check it out.' That's what led me to JGI. I'd never heard of JGI before that. I'd never heard of Jane Goodall, which I am quite embarrassed to say, considering I'm Kenyan, and she worked in Kenya with Louis Leakey. There's such a rich [Jane Goodall] history in Kenya."

That period of the new century marked Jane's official entry into the wildlife profession. After traveling to Kenya in 1957, she contacted Louis Leakey, an established and respected archaeologist and paleoanthropologist, to discuss her passion for wildlife. Jane, recruited as the secretary for Louis, proved to have the traits he was looking for to pursue his research into the connectedness of apes and modern humans. Jane's passion, attention to detail, and perhaps most importantly, unbiased mind, free from the imprint of formal scientific training, offered the fresh set of eyes Louis wanted for observational research into Tanzania's chimpanzees.

"Oh, my goodness, how did I not know about her?" Alice recalls thinking at the time. She reached out to JGI, got an internship, and stayed there for about six months, working on her research. "I was looking at Tacare as it was back then, a project, and seeing how the team had worked on integrating gender, comparing it with other community-centered conservation projects, like those in Zimbabwe and South Africa, which at the time were integrated conservation and development projects (ICDPs). It was all about finding out, 'Is this project doing enough and can we do more?' Of course, we could do more. That's what led me to work for JGI. Once I joined, I thought, 'Wow, this is really cool.' It was exactly what I'd always wanted to do, really seeing how individuals and entire communities can become part and parcel of conservation decision-making, drive conservation outcomes, even drive research that actually leads to some form of adaptive decision-making."

When Alice finished her internship and completed her research paper, Keith Brown was hired as the first executive vice president for Africa Programs. His task was to develop a strategy that encompassed everything JGI had been doing in the different African countries, which, until then, had been a series of single projects, each with its own funding. When Keith started at JGI, he hired Alice to focus on the East Africa programs, mainly in Tanzania but also in Uganda. "Since that time," she says, "I've really seen how the Tacare project evolved into the phase where all these different elements started coming into it, like human health, school scholarships, and so on. Then finally to this point where we've proven our support of the communities. Now they trust us because we have responded to their needs." When communities' short-term needs are being addressed, they can begin to think about things in the long term.

Part of the development aspect of Tacare is engaging the community members on social issues that may create cycles of repetition that are counterproductive for improving their circumstances. For Alice, one such issue is the role of women in African society. However, in rural Africa, gender roles differ not just from country to country, but regionally, and even between neighboring villages. Typically, the more isolated a community is, the more particular its way of life. Often shaped by a local religion, gender roles are a complex dynamic that greatly influence a community's future. Negotiating this dynamic can be tricky, as upsetting the societal status quo can jeopardize a community's willingness to invite JGI involvement. Although it's a sensitive topic, Alice directs much of her effort into shifting the gender dynamic in ways that produce positive results without undermining the region's cultural integrity.

As an example of these challenges, she describes visiting a village in the DRC, where she spent time with the women, looking at some water points that had been built for the community. "There was a group of men sitting and talking," Alice recalls, "and I asked, 'What's going on there?' They said, 'Oh, that's a meeting with the representatives from the village.' There was not a single woman. I asked why and was told that women aren't allowed. 'How do you communicate your needs or your concerns?' I inquired, and the women said it must go through the men." Their response was a moment of clarity for Alice, illustrating how in such communities women's needs may not be considered a priority in village meetings.

"Early on, perhaps what we could have done more of was look for more female champions," Alice says. "That's what I was really interested in—the gender dimension and seeing how we can bring in more young girls, more young women to really be at the center in terms of leading some of these activities. The more you can see yourself in others, the more

Community members gather with JGI in Zashe village, north of Gombe, to discuss the institute's integrated approach to conservation through interventions in population, health, and the environment. Alice Macharia is in the front row, second from right. *The Jane Goodall Institute, Anthony Collins.*

you're inclined to concentrate your focus. Now of course, there's a lot going on in Tanzania and other places with regards to girls coming forward, from the scholarship program and Roots & Shoots, and being at the forefront as active participants. I really wish that would have happened even more in the early stages. From the research perspective, yes, we have had many women who've come out of Gombe, but not African researchers.

"That's one of the things I'm constantly working towards," she says. "Bringing in and engaging more African women as scientists and making the conditions conducive." Nor can the remoteness of the area be used any longer as an excuse for excluding women, she says, because many Western women have come to Gombe and completed their research successfully.

The lack of women African professionals is just one result of a complex interplay of societal norms in rural Africa. In many communities, impoverished families lack the resources to send their children to school. This situation may arise because doing so would leave the mother alone to manage the laborious tasks of gathering food, water, and firewood, which can consume an entire day. Even if funds are available for schooling, the housekeeping mother cannot manage her youngest children and household chores without the help of her school-aged kids. Likewise, the father of the family may require assistance from his children to maintain small subsistence crops or to help him during the long days of fishing.

Essentially, many school-aged children are necessary team members in the work of securing daily survival for the family. Even if funds and time are available for one or two children to attend school, the preference is almost always given to a boy's education. While this preference may reflect the values of a patriarchal society, it's also because young girls are often considered a poor investment where education is concerned. The reason? Puberty.

In the Hoima district of western Uganda, Yakaka Saweya is the deputy head teacher and senior teacher of female students, known as a women's patron, at Wambabya Primary School. Mrs. Saweya, as she is respectfully called, says, "I coordinate peer educators in the Kikuube district and Hoima district. I'm in charge of taking care of the girls, especially during puberty, coaching them on why they shouldn't drop out of school, and how to manage themselves during their menstrual periods. My role is to keep them in school and to also help them avoid early pregnancies so they can finish primary level and go on to secondary education. At least then they may end up reaching a tertiary level. In the past, almost every girl's education would end at the primary level."

For cultural and social reasons, this topic of puberty and menstruation is one that excludes men. Unlike the Roots & Shoots approach, local educators need to be at the forefront of this gender-specific, culturally sensitive issue—not only because it affects the school curriculum but also because the parents need to be involved. Where an NGO would be limited by an individual's secrecy, shame, or resistance, local educators hold a key for unlocking a potentially taboo or unmentionable topic.

"We organize information or sensitization meetings in different communities," Mrs. Saweya says, "to sensitize the parents about what we are going to do. We discuss the aims of the program, to look at the girl children of the household and how their future studies can be accommodated. The parents need to understand why the program is in their community, which is because girls are dropping out of schools at a very high rate. In the beginning, we had ideas but nowhere to begin from. When JGI came, their research found that most girls were missing schools during menstrual days. They couldn't manage themselves during these days because they had no sanitation materials to use."

Mrs. Saweya explains, "If they're in this cycle for five to seven days, that means they will miss school for all of those days! Now multiply this by three, the months in a term, and sometimes add the days when a girl feels weak. All these absent days contributed to poor performance from our girls. This is when [JGI] came up with the idea of using reusable sanitary pads. At first, they started providing these to schools where JGI was already involved, and these washable pads can last a year if well looked after."

Over time, Mrs. Saweya explains, it became apparent that the initiative was too limited in scope and too expensive. The girls needed to find a creative solution. So, the schools asked JGI if they could send people to train the girls and teachers to make the pads themselves. "JGI had to look for the right person," she says, giving a dismissive hand gesture as if this took far too long for her liking, "and they then brought the resources—the materials to use. The girls were trained on how to make these sanitary pads, and now most girls in upper classes can make these sanitary pads by themselves."

She adds, "Not only girls, but also their parents, because as we teach the girls, they train their own mothers. They all need them, and they are of much benefit to a household. They're not as expensive as those sold in shops, which can only be used once and just thrown in a pit latrine, or sometimes they just throw it anywhere in the environment. We are encouraging the reusable sanitary pads, and if we get the means of a sewing machine, we can add quality to these ones made by the girls and their parents." She predicts that the reusable pads, which are better for the environment, eventually will supplant the disposable ones.

Something as basic as meeting feminine hygiene needs can have a profound impact on the future of women in skilled positions. "When you compare their performance now," Mrs. Saweya says proudly, "the girls are performing better, and are succeeding. Before JGI came in, these girls would miss class on every single menstrual day. Previously, my school used to perform poorly, and the girls used to drop out. Since this initiative came in, my school is performing better, and among the children who sat exams last year, *all of my girls passed.*" She punctuates the last five words with a forceful index finger on the cover of her leather-bound notebook. "They all passed, and they are now ready to go onto secondary school. If more girls finish school, I look at their future being bright."

Echoing Mama Sania, Mrs. Saweya describes the challenges facing local communities in which many people have not completed their education and opt to produce as many children as possible. "At the end of it all," she says, "they can't care for those children because they're unable to afford their schooling, they can't afford medication, and they can't even afford clothing." Now that the girls in her school are completing their studies, she says, "we're focused on the villages that are behind, that don't have ideas on how to protect their girls' future. When we see that the community has taken this initiative, then we move to another. So many are coming to ask, 'How are you doing this?' When we tell them, 'We have an NGO which is helping us create mass awareness,' they always ask how to become a part of it. We want to be examples for the neighboring areas, and I'm sure even when JGI goes, this will continue. We want it to be sustainable."

For Mrs. Saweya, this work is crucial to the communities, because if they can keep their girls in school, they will be achieving something significant for the future. "Now," she says, "the only challenge left is that the pad is made freehand—it isn't the same quality as one made with a sewing machine. We have the challenge of getting them a sewing machine. Then again, with training, we could impart the skills of using machines to these girls. Not only can the girls then make higher quality pads, but they then will have the skill of machine sewing. This might help them generate income on their own."

The need for such basic resources to enable teenage girls in Africa to attend school is only one of the hurdles they face on the road to skill-based employment. Another is that male and female roles are still more strictly delineated in African nations than in many Western societies, a lingering set of cultural norms connecting their traditional way of living to today's world.

"Culturally," Alice says, "my own education and thesis research was a big thing. I think it's still a big thing because, as a woman, you're going into the forest, you're going away from home. I remember when I was doing my research and going to the Tana River, my mom questioned it, asking, 'What are you going to do there?' Everyone in my family did, asking, 'What, where is that?' I exclaimed, 'My goodness, it's in Kenya.' So, there's that cultural hesitation, those barriers that prevent women from participating. The more we are aware of them, the more we can address them, and the more we can create environments that facilitate participation and full involvement of women, the better for us."

Alice adds that she finds the conservation community still driven by men. "Especially in parts of Africa," she says, "in many places you'll attend a meeting, look around, and realize most of the delegates are men. For young girls, I think it's important early on that you ensure ways of keeping that pathway open, creating an environment that gives them that choice. In Tanzania, for example, in recent years, most of the Roots & Shoots members have been girls. Then how do you ensure that that same interest, that same enthusiasm continues as they go into high school and then as they choose their career paths? Then afterwards, how can we make sure that the excitement and enthusiasm continues, and that it's not necessarily either influenced by culture or downtrodden by other people?"

Alice answers her own question with an example from South Africa. "I remember a great example was the WASH program (water, sanitation and hygiene) in South Africa, where in some places it [gender discrimination] is even worse than where we are working. The way they approached gender inclusion was by training the women in a technical skill, so that they became the technical expert. They had to be invited into meetings to speak as the skill person. Ordinarily, if you had a boy and a girl, who would you send to school? You'd send the boy to school. Yet here, you are seeing this woman or young girl talking about fixing a well or designing infrastructure for a sanitation area, talking about something knowledgeably. It starts changing how others view some of these long-standing cultural issues and beliefs.

"This is the way for people to start seeing women as contributors, as being intelligent, and as being worthy of investment. They wouldn't ordinarily invite a woman, but because she is conversant in a specific topic, she's shown her relevance." Men begin to realize that the women are educated and have much to offer. And seeing women involved prompts other women to question how they can also be included and be seen in the community as contributors and thinkers, as providers of information and solutions.

As the concept of women contributing to community development takes hold, NGOs can ensure that the momentum doesn't stall by preparing young girls to understand the scope of their future opportunities. The Roots & Shoots program, for instance, specifically invites a 50 percent participation rate of female students, and the JGI Girls Scholarship Project provides education opportunities. Meanwhile, as an African and a woman in a senior professional role, Alice is well positioned to instigate change. With a deep understanding of context and a nonconfrontational approach, she and other such architects of the social future can pave the way until the next generation makes the change more systemic. "Otherwise," Alice says, "those opportunities may never come."

Chapter 17

The cycle of regeneration

Alice Macharia is in it for the long term—and the short term

Imagine a bare hillside, stripped down to nothing but the eroding topsoil. Natural regeneration begins with seed recruitment, and grass seeds blown in from nearby areas thrive because the vacant soil is devoid of competition. The sheer volume of grass roots stabilizes the topsoil, reducing erosion, and begins the cycle of fixing carbon and depositing nutrients. This cycle opens the door for herbaceous plants such as many weed species, which are transients that move between disturbed areas. Enriched by the stable, nutrient-rich soil, they also thrive, growing much taller than the grass to create a protective microhabitat underneath their foliage. Under the decay of herbaceous deadfall, microorganisms thrive, preparing the soil for the next stage of larger plant forms.

Until now, these species have been fast-growing and rapidly reproductive. Next, protected by the cover of weed and fed by the healthier soil, woody shrubs begin to emerge. By the time these shrubs break through the height of the herbaceous plants, they have enough stability in the soil and strong enough stems that they no longer require protection. Shaded by the now much larger shrubs, the herbaceous plants begin to die back, throwing one last seeding event for the winds to carry the next generation to a distant vacancy.

Under the stabilized and sheltered conditions of the woody shrubs, tree saplings begin to appear. The lifespan of the shrubs ensures that the trees have the protection they need in their early years, enough time to establish their own home on the hillside, growing the largest root mass of them all. Central stems shoot above the height of the shrubbery until the trees capture most of the sun's energy. Growing taller, wider, and living for decades, the long-term strategists have arrived, and a restored forested habitat takes form.

For any NGO working in developing areas, this cycle of regeneration offers an important lesson. Long-term and sustainable capacity-building doesn't happen by itself. It takes many years and is supported by a sequence of smaller, faster growing initiatives that stabilize the foundation of change and enrich its soils with trust. Alice Macharia, vice president of Africa Programs for JGI, has nurtured this cycle for many years in many different contexts.

When Tacare first started working in the DRC, she says, they had to assess the successes and some of the challenges of the Tacare project in Tanzania and then decide if it could be replicated in the context of the DRC. From an assessment published in 2002, they knew that the community aspect of the program was key. Yet, in the DRC, the communities in question were living under the near-constant shadow of conflict and war.

"I remember one visit which, as usual, was a logistical challenge even getting to the DRC because of insecurity," Alice recalls. "As we're taking this journey, I was thinking, 'Wow, this will be difficult,' because you need a conducive environment for people to be active and participating, and not always worrying about their own security. For example, I could say something that might ignite a conflict that I didn't even know was brewing, or we could be working on an intervention and then in one instant, everything just disappears. All the work that you've been doing with the communities, all the close discussions, them selecting who's going to be the local resource people—you go through that whole process and it could just end.

"Right when you feel like you're getting headway or you're moving in the right direction, something happens, and you have to stop the program. The DRC is one of those places where you often wondered, 'Will this even work?' One moment you're able to continue because you have a really good team, then the next moment the work stalls because the local resource people on the ground pack up and go."

Alice notes another important contextual difference—that conversations with the villagers were always about the difficulties of living in the DRC. "One of the things you usually notice when you go to a village is that it's vibrant, there's chickens running around and there's just stuff happening," she says. "When I asked where all the activity was, where the animals were, the women said, 'Because we've been through this [conflict] period, we just stopped keeping these things. The rebels would just come and take them.' Anything they had that seemed to be of any value was taken away."

Under the boot print of war and decades-long civil conflict, livestock is one of the first local resources plundered in these often protein-starved areas. As Alice and her team listened to the community, they realized how challenging their task would be, even if they moved slowly. "But we could at least keep engaging the people, find ways to bring women to the table, and keep having these conversations. People hadn't kept livestock in a really long time," she explains, "but it was one of the things they highlighted. We wanted to conduct a pilot project to see if once again having subsistence livestock at home would satisfy as an alternative to bushmeat consumption.

"There were so many things that went wrong with that initial batch of livestock in the project," Alice says, reflecting on the project's initial missteps. "A big lesson learned was regarding environmental exposure and using animals from outside the area. They didn't do well because they couldn't adapt to the local climate. So then you have to take a step back and understand how to think through these initiatives a little bit more and ensure that it's not only culturally, financially, and environmentally viable, but that it makes sense to be doing it at that moment in time. The project itself may make sense, but if the timing is off, the communities might not be ready to get involved."

Another consideration is identifying the right community partners—other outside groups with which to align JGI initiatives. "Going in, you may think they're really community driven, they sound like the perfect partner who's going to help us drive change. But if we discover that someone could be using the partnership in a way that is not meaningful, and not how JGI wants to work, we need to reconsider that and end it."

Alice points out that the same standard applies to JGI's own initiatives—for example, an early palm oil project. "We needed to step away from that," she says. "It was one of the activities in the past that JGI was working to implement as a sustainable, small-scale subsistence crop using hybrids. One local company then started buying all these seedlings and taking them south, where a big farm was established. It didn't last very long because we backed away. It was taken from a subsistence level and upscaled to a commercial business, the impacts of which had the potential of harming the riverine forests."

When JGI opts to step away from a partnership or from its own program, the organization might seem fickle or irresponsible, but it's an important mitigation tactic that ensures the core principles and values of Tacare are not diluted over time. The JGI Tacare Bill of Rights guarantees that communities will be treated with respect, dignity, and integrity; maintain their right to self-governance; and participate with JGI voluntarily on a basis of open dialogue and mutual trust. Merging with an organization that does not support the holistic approach with complete transparency, satisfy the Tacare Bill of Rights, or serve the broader goals of the institute could compromise JGI's local relationships some 60 years in the making. Knowing when to step back and reduce JGI involvement or halt a misdirected initiative is a critical attribute of the approach. Sometimes, stepping back results in a forward movement.

"I remember one village meeting," Alice reflects. "We arrived expecting to attend, but they told us, 'No, we don't want you here. We don't

JGI has sponsored billboard signage in the DRC. These signs are posted in marketplaces to educate locals about bushmeat consumption and live animal trafficking. *The Jane Goodall Institute, Fernando Turmo.*

want JGI involved at all.' In this village meeting, it was an open JGI forum, one where people are supposed to speak their minds. Well, they spoke all right, repeating, 'No, we don't want JGI involved.' In such an instance, even gentle persuasion or negotiating for inclusion, regardless of the potential short-term benefits, could be considered strong-arming for an outside agenda. Our response was 'Sure, no problem. When you are ready, or if you would like our input, please call us.' At that time the village government was at the center of discussions. Information surfaced, we later found out, during the question and answer session; they were not necessarily against JGI—they were actually against the community leadership that hadn't honored something, but because this was a JGI forum that brought the community together, they rejected everything.

"After that meeting, Emmanuel Mtiti [director of programs and policy for JGI-Tanzania] explained everything, saying, 'Yes, we're actually going to stop working and whenever they're ready, they know where our offices are,' and he told them, 'Please send your representative when you're ready to continue working with us.' It was tense. Oh, my goodness, was it tense! The people were furious at the village government. I'm not sure if money had been directed towards things it wasn't supposed to be used for and they were not being transparent about it."

Yet, about six to eight months later, the villagers requested JGI to return and start working at the village level again; apparently, the village government had sorted out its issues. "It was really good to see that our involvement was not being forced upon them," Alice says. "Because if the community has rejected it, for sure, there's no point in moving ahead—because whatever you propose, especially at that time, will not be adopted by the community. Just give them the time, the space to resolve what they need to resolve, and know that you will be there at such a time when they are ready to re-engage."

As Alice makes clear, these moments of resistance can be an important opportunity to nurture trust if handled in a respectful way. Also, a proven history of working through problems and returning to an amicable partnership suggests that any future issues are not insurmountable, creating a sense of safety in the relationship. Just as a tree adds rings of growth in response to favorable conditions, so trust grows more robust over time as it becomes layered with a history of positive outcomes.

Though long-term involvement generates meaningful change, Alice insists that short-term involvement is important too. "Conservation is something so long-term," she explains, "that if the local people aren't seeing a short-term benefit, then they won't become invested in the programs that require a longer timeline." Deciding which of these strategies makes the most sense for a given initiative comes down to an NGO's ability to listen and understand.

Throughout the natural world, plant and animal species also must strategize to endure. To survive, they must be successful at achieving space, nutrition, and reproduction, employing innumerable strategies to attain these ends. Consciously or not, everything a plant or animal does in life is geared somehow toward successfully crossing nature's finish line, reproduction, and some of the most spectacular events in nature reflect this unstoppable quest. Think of the great wildebeest migration of East Africa's Serengeti and Masai Mara, with as many as 1.5 million of these bovids on the move; the 500,000-odd sandhill cranes making a Platte River pit stop as they migrate toward the Arctic Circle; the whale highway off the east coast of Australia where humpbacks complete a round trip of 6,000 miles (10,000 kilometers) before returning to the Antarctic; or the periodical cicadas, some which live for 17 years beneath the earth's surface for the sole purpose of emerging en masse to mate. These epic events all revolve around sending forth the next generation, ensuring that the species survives.

Emphasizing the importance of choosing which strategy best suits a community initiative, Alice says, "I think that understanding timelines is critical. It gives you an opportunity to step back, perhaps even start over, be able to learn, and where needed, circle back with improvements. A typical [short] funding period for an NGO initiative is a start, but by the time you begin making improvements, the project is over, the funding is gone. Everything that you basically worked to address goes backwards and the project is finished. For these communities, knowing, understanding, and believing that JGI is going to be there into the future helps them feel secure in their own capacity building."

Even if a predicted timeline suits the project, setbacks are common, and adaptability becomes the next key attribute. "One thing I must say," Alice says with a smile in her voice, "is that because things are changing and people are talking to each other, we're able to make those adaptations as we're moving along. For example, our land-use planning work. We're progressing well on the greater Gombe ecosystem." She explains how Lilian Pintea created maps with inputs from local communities, scientists, and other stakeholders to envision a possible wildlife corridor that also restores watersheds and protects people from landslides and flash floods. After going through the process of discussion and sensitization, the local communities set that land aside. But sometimes, later on, those villages would be subdivided. "Then what happens? All the work that you did will need to be redone."

Yet, says Alice, because of the long-standing relationships developed through the Tacare approach, some of the communities continue to honor those land-use plans, and the areas put aside as forests are still being protected. "I think those are the moments where there's the potential for things to really disintegrate. But because you've taken the time, the local communities really understand why some of these things are happening; they continue to support those processes." Alice believes this is because of JGI's Tacare approach and its investment in developing and training the land-use management teams, the village champions of conservation, and the village leaders.

"So, the teams are already in place, already patrolling the forest and continuously educating people, even when the village is [sub]divided. If it's divided, say in half, perhaps some of the members of one team end up in the next village—they still want to continue with that same ethos. They continue to push for this work." For Alice, the forest monitors are crucial, because they are constantly collecting data upon which the village government can base its decisions: "I think that's the critical factor in terms of having a core group of people who are completely bought into the idea and will continue to be champions even after the project ends."

Whereas community-led conservation initiatives can be rapid or slow-growing, the Roots & Shoots program is closer to a long-term strategy since the education of today's youth is an investment in tomorrow's decision-makers. Sharing an encouraging story, Alice recounts how, many years ago, there was funding for a program called the Environmental Education Project in the eastern part of Tanzania. "We were going on a monitoring trip," she recalls, "and driving just on the outskirts of Dar es Salaam, when we get flagged down by a policeman. Of course, immediately we wondered, 'Oh, what's going on? Why have we been stopped?'

"The policeman came to the side of the vehicle and said, 'Do you know why I stopped this car?' The driver said, 'No, because I don't think I was speeding.' 'I stopped it because I saw Jane Goodall Institute and Roots & Shoots [on the van],' explained the policeman. 'I was a Roots & Shoots member in Kigoma. This spot I'm sitting in is the spot I chose because of that tree,' he explained while pointing. 'It's a beautiful, fun tree, and there's all these birds and nests about there. I chose it because it reminds me of all the things I did when I was in Roots & Shoots in Kigoma, working in nature and learning about the environment. Sitting under this tree, yes, it's part of my job, but it also makes me feel at home.' You might say that a boy from rural western Tanzania, growing up and becoming a police officer in the nation's far east and still appreciating those Roots & Shoots experiences, is a good example of how the right lessons at the right time can stay with a child well into adulthood."

Alice offers another, more recent example from Bubango, in Kigoma district, when she and her team visited a community that had been working with beehives. During a meeting in town, the locals showed them everything they'd been doing with the honey and the training they'd received from Tacare. They were even developing their own packaging for the honey, which displayed a conservation message. Toward the end of the meeting, one of the beekeepers asked Alice if she wanted to come and see where they kept their beehives. It was completely unplanned, but of course Alice agreed.

"We drove out with him and two other ladies and walked into this forest. It was so great to see the pride he had. It was showing me how they had completely bought into this approach, and how they had also made the decision to dedicate some of the family land to forest. A few of the other beekeepers were also putting their beehives in this forest and had set up a place outside with benches where Roots & Shoots groups could come and just feel the breeze of the forest as they're taught about the environment. I thought that was great. It was just everything coming together in terms of a community completely adopting the idea, demonstrating an understanding of why it's important to conserve their own forests."

As Alice's story illustrates, the beehives can serve as a catalyst for local people to better understand the connectedness of humans and the environment and how sustainable resources can serve both sides. The bees provide honey and income to the villagers but produce poorly if the nearby forests and woodlands are compromised. In essence, bees provide a bridge between the needs of an ecosystem and the needs of the people. "You just build on that," Alice says, "and keep working and working to improve both their livelihoods and natural resources. The local people see why it's important to protect them, while also benefiting"—benefiting, in other words, from both strategies, the long- and short-term approaches to conservation.

Chapter 18

A 'talking office' with maps

Joseline Nyangoma wants science to tell a story

Compared with written or verbal information, images seem to be the fastest way for human brains to process large amounts of detail at once. Imagery also does a good job of communicating time and space. As Dr. Lilian Pintea explained earlier in the book, GIS maps and satellite imagery capture precise spatial and temporal parameters that can serve as a common language in documenting and transferring knowledge to or from local communities. Not only is this common language a more objective and transparent way to communicate such knowledge, but shared access to imagery increases the meaning and ownership of information and people's trust in information to guide local decisions.

As head of the Department of Natural Resources in Uganda's Hoima district, Joseline Nyangoma understands all too well the importance and power of imagery, especially in the government sector. A few GIS maps and satellite images in an otherwise mundane bureaucratic document can allow government decision-makers to quickly visualize and thus grasp the key points about something such as habitat health. To influence government policy and bridge information gaps, Joseline is advocating that the real-time data collected by forest monitors be merged with other

Joseline Nyangoma. *The Jane Goodall Institute. Adam Bean.*

information and data sources in GIS maps. Yet, despite her best efforts, this is a goal she has yet to achieve.

Joseline began working for the Department of Natural Resources in 2000 as an environment officer and later became a senior environment officer. Now, she is head of the department, which oversees forestry, wetlands, and community lands. Her district, Hoima, in the midwestern part of the country, administers five subcounties and is responsible for guiding the conservation and management of natural resources in the district. According to Joseline, wetlands and forests are the key areas being degraded, and most of these are found on community land. Although her department conducts periodic wetland inventories, she says, "under forestry, we have more challenges than our budget allows us to manage, so we must be selective with our activities, and thankfully, NGOs come in to help us with some of those issues. That is how we became involved with JGI, which came to help us with community landscape challenges."

Joseline speaks passionately as a public servant who cares deeply about her work for the environment. "Most of our forest loss," she says, "is due to poor farming methods and forest fire, and when you look at those two areas in terms of biodiversity loss, I think we've had more than we know, because collecting and updating our data is always a limitation. There were funds allocated, but now we rely on local governments for collecting this habitat information, especially those areas that are private forest.

"With our limited budget, NGOs have been assisting by developing land-use plans, and we facilitate community sensitization to boost the program. JGI and similar organizations helped us with the formation of forest monitors, who are collecting daily habitat data, which is good information for us to access. The challenge that comes between us and the use of these data is departmental structure." Because her department operates at the district level, she explains, it lacks the capacity to send staff into local communities. "That is why I'm advocating for these forest monitors, because I see they have the information we lack." That information flows to community decision-makers with support from the NGOs but not yet officially to her department.

Joseline is interested in understanding how and why those communities have been able to keep focusing on their forest reserves for so many years, especially when the members don't always see direct benefits. But she believes that, through the forest monitors and community sensitization, residents began to understand the value of the forest. In addition, she says, they have been taught about other income-generating activities or enterprises that can be developed around forests, such as beekeeping and tourism—activities that conservationists continue to advocate so that local communities preserve the forest lands against competing farming practices, such as large-scale sugar cane or tobacco production. "We have been talking about carbon storage," Joseline adds. "However, some people don't understand it." Despite this challenge, she believes that "when you work from bottom up, from the communities, then bring it upwards to schools in environmental education, then conservation can be achieved."

For this reason, Joseline wants to support communities that have formed associations for private forest owners, because they deploy forest monitors, who are valuable assets as citizen scientists. Because monitors can accurately describe what they observe on the ground using smartphones and mobile apps such as ArcGIS Survey123, Open Data Kit, and Forest Watcher, their findings can contribute significantly to local conservation efforts. "It's much easier to achieve conservation in a Central Forest Reserve because there are no human beings," Joseline says. "But remember, we are protecting those forests for the community, on community land, which is difficult when you consider how they're using this land. When

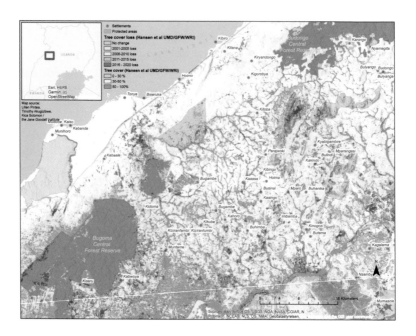

Protected areas (cross-hatch polygons), remaining tree cover (shades of green), and tree cover loss (shades of purple) between 2000 and 2020 in the Budongo-Bugoma region of western Uganda. The darkest shade of purple shows the most recent tree cover loss (2016 to 2020). *The Jane Goodall Institute, Lilian Pintea.*

you look at the initial map classification of these forests," she says, pointing to a large map on the wall next to her desk, "about 70 percent was private forests and 30 percent was the protected forests in reserves back in 1990. But when you look at 2015, you can see it's the reverse, and the protected forests in Central Forest Reserves are now at 70 percent and the private forests are being lost, down to about 30 percent.

"Most of our areas here were characterized as tropical high forests," she continues, "and people would sit in abundance—they would never imagine it could all disappear from being converted to agriculture. But people started clearing because they would always say, 'But we don't want monkeys. For us, we want to cultivate.' So, they couldn't see the benefit.

There's a document we developed, a strategic plan, showing the forest corridors and how they needed to be preserved. I was quarreling with another big NGO about how they were going to protect these corridors, because it was all focused on the forests and not the people. So, with those organizations, it didn't work, and they retreated to just working in the protected reserves.

"Now, we have an updated map." Once again, Joseline points to a map on the wall. "The dark-green forests are gone, and it now has more yellow and red, which is cleared land. It won't be until we go back and work with the communities that we can improve on the corridors." Her goal is to keep supporting the community forests and the community forest monitors, working with them, even bringing them on board as part of local government. She acknowledges that the current system doesn't allow for this. "But you never know, maybe in the years to come when we have a government restructure, those are the things that should be recommended."

Joseline is candid about the bureaucratic constraints she faces. "Our structure is such that we have multiple positions for protected reserves, but for all the remaining land like wetlands and forests, we only have two. So, you can see, we don't have the resources. We are supposed to have a forest committee down there working with the communities, but again, funding is not allocated. I think it's also because we all want to work like islands: the environment people are here, the lands people are there, different NGOs elsewhere, all of us are scattered. Bureaucracy separates everybody, and possession over knowledge, ownership of 'results,' separates everybody. Somehow, we need to start at the beginning and become a team."

That team, Joseline believes, should include community members such as the forest monitors because they are capable of collecting the necessary information. The monitors are trained, on the ground, equipped with smartphones and apps, and already collecting basic data that is relevant and contextualized. "If we improve," Joseline says, "with this new technology, you could tell a monitor down there, 'Take your phones and collect this and this.' Then they can forward the data to us or others. We can achieve so much more besides just looking at encroachers. Sure, find where people have come and cleared—they can report that, and we will come in for enforcement. But these forest monitors could also look at biodiversity because they're in there every day."

According to Joseline, outside organizations are unfamiliar with all the biodiversity in these areas. They may have some information about the medicinal plants because people are actively seeking them, but they know little about the animals. Yet, she says, "When you ask those community monitors what animal species are there, they can tell you. I don't want to hire university students to come and do the work—I'd rather they come and help us facilitate the process and work with those communities. Sure, these students have set up questionnaires, and we can get information for conservation, environmental resources very fast, but we are lacking it because we don't have the connectivity down in the community." Again, Joseline expresses her frustration at her department's current lack of capacity to integrate the data from the forest monitors' apps into its own system.

Once they have that capacity, Joseline says they can gather the information and produce a GIS map. And once the data is represented in a map, it can be presented in council meetings, so that she can tell the members, "Look, this is the state of the forest." Since some of the councillors have had limited formal education, images and maps are often the most effective means of communication. "I can present a three-year report to them just by showing a map that has data on biodiversity, on the wetlands, on the forest, et cetera, to see how our natural resources are changing," she says. "I'm very sure that even the National Environment Management Authority would be happy getting access to this kind of consolidated insight. These other organizations now, like the consultants who are always looking for information, can also use these data. Because if we're honest about our biggest problem in this department—it's access to information!"

With a map, Joseline says, you can guide people visually during a meeting, beginning with the forest. "From there, you take them into the wetland. And sometimes when you find a forest, you might even find a wetland inside, or some of them are more of a riverine forest. When we begin reporting in that form and you really show them the source of data, how it was collected by the community members, then maybe the sub-counties can begin budgeting for such programs. Then from there our government will realize that we really need them."

The wheels of any government turn slowly, and Joseline is on a mission to have the relevant agencies legitimize forest monitors by incorporating their data into GIS and developing maps to influence policy. At the government level, Joseline says, a request to recruit local staff would be seen as problematic. But community monitors could be incentivized to do this work on a part-time basis, and "in the generations to come, they will be part of those communities, so they'll stay there. These can become normal positions within a village that are occupied by those working on behalf of their own village, and for sustainable outcomes. I want them to become natural resource managers down there, so that when it comes to land issues, they can tell you; when it comes to the wetland issues, they can tell you. Soon, you'll find they have mastered a large range of different biological knowledge relating to plants and animals. In the future, this will also help us to incite our communities to promote tourism."

For instance, Joseline says, she once visited Murchison Falls National Park and went with the locals to an area where pythons were said to be. "They told us, 'Those pythons, they come at this time.' We went there and we were waiting; when the sunshine came out to the right spot, we saw the pythons coming out like it was planned. You get it? It is those communities that have this information, and we're not capturing it or using it for conservation purposes when we really should be. Then there is also the historical knowledge. When you engage with them and ask them, 'Do we have this?' they can teach us. If we get time in those communities, they can teach you a lot of things you didn't know.

"We can approach even pastoralists, the cattle keepers," she continues, passionately. "They are very, very useful. Even when it comes to things like medicinal plants, they know those things because they know what is good and what is not. Then we can talk to those who collect water and ask what is changing in their daily landscape. I have a list of people I think we should talk to, and I think they're all useful at the end of the day."

As a possible model for the formal integration of local forest monitors, Joseline points to the village health teams (VHTs), which are recognized under the Department of Health. She wants to study how those teams came to be recognized, so that her department can apply those lessons to establishing its own partnership with the communities. Joseline also emphasizes the need to organize local communities by doing a cost-benefit analysis so they can begin to understand the value of conserving the forest and the wildlife and also see how the landowners will benefit. "Specifically, we want landowners with small plots to benefit. If you compare the people who have small pieces of land with those who have large ones, the people with less land could benefit far better if they chose the right land use. The large plots of land are more difficult to manage, so they usually get used for large monocrops, which don't really benefit the local people." The goal is to educate landowners large and small in conservation farming.

Little by little, Joseline believes that larger crops and huge amounts of agricultural land will be eliminated. People have already started seeing the downside, she says. "For example, the youth like growing tobacco. But they could spend a whole year on tobacco and get nothing out of it. I always do the cost-benefit analysis, and I tell them, 'You spent the whole year when you were planting tobacco from now up to December. You earn once—the little money you're getting while keeping yourself and your families malnourished. All that time, you are all in the hospital. Then there's your school fees, and at the end of the day, there's nothing, and you're not getting rich. Do you think that is the best?' People have started seeing that, as a crop, tobacco doesn't work here."

A sugarcane plantation outside Masindi, Uganda. *The Jane Goodall Institute, Shawn Sweeney.*

Speaking with even greater conviction, Joseline says, "I want to spend time with my communities because even if I don't have GIS maps like these here," on the office wall, "I just draw. I look for markers. I tell my community, draw the maps. Even though all of this structuring will take time, I believe we can get the forests back. I believe so because for us, when you look at our climate and environment here in the west of Uganda, these rains give restoration and, if given a chance, the forest returns quickly. Things will restore, then maybe we can even add enrichment planting because we have the seedlings which we are using to teach them [local communities] about collecting and propagation. The other day, I was teaching them how to pick wildlings, the ones that have germinated."

Joseline's dream is to begin mapping the forest—not only for purposes of conservation but because, she says, "the other big issue down there refers to land title, and I'd like to come up with some sort of customary certification. Customary certificates of ownership, where an individual gains a real land title, which should reduce the land disputes." She speculates about offering land titles to the forest monitors as an incentive, but, on the other hand, "the last time JGI was trying to do that, most people were stealing other people's land, they were going in our riverine forest." Proper documentation might solve that problem, Joseline suggests, but since her department doesn't have feedback from previous efforts, it's hard to apply any lessons learned. "It's like I mentioned before," she says, a tone of frustration creeping into her voice, "we need systems for both keeping and transferring information."

As Joseline points out, "Between the village communities, the government sections, organizations like JGI, and other NGOs or universities, we should have a wealth of information at our fingertips." And the thought of all that unshared and inaccessible information brings her back to her favorite topic: maps. "That's why I love these GIS maps," she says, "because these can be shared resources. I tell people I want to have a talking office. When people come here, I want them to see my sub-county on the walls. I want to transform this same space into one that shows you the information just by looking at wall-to-wall maps." A map, Joseline declares, "is something that can talk for itself. I want our science in picture form so that we can tell a story."

Chapter 19

People, pixels, and puff adders

Lilian Pintea contemplates different ways of knowing

A person's perception of the world around them and their place within it is a complex system, an ontology that is neither true nor false but a reflection of their community's cultural, environmental, and spiritual beliefs. This consideration is important in the countries where JGI operates. Tacare practitioners work to remain adaptive in their approach, adjusting it according to the needs and customs of the people of the place while supporting their efforts to create positive change in their communities. In such situations, visitors must build a fuller picture of the local values and world view to create a foundation of trust. Yet there are times when much is lost in translation, despite good intentions. Knowing how to work across different cultures and disciplines and how to incorporate different ways of knowing has been an essential element of Dr. Lilian Pintea's work with the Tacare approach.

There will be times when patterns shown on a satellite image can only be explained by asking the local people for context and meaning, Lilian says. "Satellite imagery is great to help us see where things are. This is important when we try to understand the status and trends of forests, land use, and so on. But, though the satellite map can show us that a patch of forest is there or how it's changed, it can't always tell us why."

For example, Lilian says, at the end of one of the first participatory mapping exercises in the Gombe region, looking at the satellite image overlay with people's notes, he asked, "Is there anything missing that you think is important and should be on your community map?" In some villages, participants decided to map their traditional sacred sites. "When that first happened, I was completely unprepared!" he exclaims. "I thought, 'Oh, my goodness, what do I do? Do I support this traditional belief? What do I do with the information? Do I protect it? Do I share it?'

"Since this was the people's decision, I knew it was important information to capture on behalf of those communities who wanted to share it, so I began digitizing and entering such traditional knowledge into GIS as its own separate layer. I did not expect that later those sites would explain so much of what I had been seeing on satellite imagery as I was trying to map and understand land cover and land use change in the Gombe region," he continues. The first insight was from a village north of Gombe: "When I first saw the satellite image, I was struck by this single patch of forest completely isolated and surrounded by farms and settlements. This forest was protected by the local communities as a Tacare project forest and was one of the first features identified on the community map. As the participants started to add traditional sacred sites, the first was a site on top of a hill. 'We used to have an old man who lived up there,' they told me. 'He's passed away now, but he used to know how to call snakes. He could talk to snakes.'

"Then another person pointed to a tree on the map. 'You see here, that tree here, this is a sacred tree. There is a spirit living in this tree.' We went on the ground and visited the tree to confirm its location with GPS. It's a beautiful tree, with stones placed around it. As we were mapping the tree with these villagers, an older man was watching us. After we left, he came and checked that the tree was OK."

Then, Lilian recalls, another man said, "'Oh, and by the way, we really want to map this large rock here in the water where the forest meets the lake.' I couldn't understand why he wanted to map the rock in the lake. Remote sensing specialists mapping forests would usually just cut those rocks out of the image—they're just noise. But he told me, 'The fishermen's job is really risky and difficult. Before they go fishing, they come to this rock for traditional ceremonies, asking spirits to allow them to be safe and catch fish.'" As Lilian shows this on the map, it's clear that these three sacred sites—the house of the old man who talked to snakes, the sacred tree, and the sacred rocks at the edge in the lake—mark almost the center and two outermost edges of the forest patch that was saved and protected by the community.

"Some may look at this and say it is a coincidence that this forest patch is still here, surrounded by the traditional sacred sites. I don't think it is a coincidence." On the contrary, Lilian believes that this and other forest areas on the village lands were protected because of their spiritual significance. "Even though the sacred sites and spirits may be not seen, they're real in many people's existence. They're real because they manifest themselves in people's behaviors or attitudes that, in turn, shape land cover and land use."

When the greater Gombe ecosystem CAP was developed in 2005, outsiders didn't include this forest patch as a core conservation goal because it was too far away from the other forest patches and chimpanzee range, Lilian says. Nor did that forest patch seem particularly significant for restoring or protecting watersheds. Yet, when community members developed their first village land use plan, they decided to preserve it on their own initiative. "It's still there," he says. "It might not be there tomorrow because the younger people's beliefs and attitudes are changing. But the hope is that communities will adopt new concepts, like land use planning, and have the resources and capacity to continue to take care of the forest as part of a healthy landscape."

Understanding local belief systems does not necessarily mean endorsing, promoting, or embracing them, Lilian says. For his purposes, they offer insight into the context of a person's or a community's relationship to their environment, while furthering respect for local values, trust, and connection across different cultures and ways of knowing.

"What I've learned from the local people in the Gombe region is that

Village forest and sacred sites are mapped by the local community. *The Jane Goodall Institute, Lilian Pintea.*

spirits are everywhere," he says. "They're in the lake, trees, stones, and streams. You don't cross a stream without greeting the spirit. When I go for a walk in the forest, I enjoy the surroundings, observing the plants, animal signs, smells, and sounds. Well, for a local person from villages in Gombe, they might greet maybe more than 10 or 15 spirits on that same walk, but you just wouldn't know, because it's not necessarily an outward ceremony—it's internal, done with their hearts and minds. It's embedded. To me it is like a ritual of constant praying. It connects you with the spirit; it connects you with the forest in a more meaningful way."

Of course, Lilian adds, this practice of greeting the spirits might have been more prevalent in the past. "These communities, like many other communities around the world, are dynamic and are changing as we speak, influenced by Christianity, Islam, the teachings of science, by what people learn in school or hear on the radio and see on television and, increasingly, on social media."

In Lilian's view, such transformation often entails a loss of precious traditional ecological knowledge and rituals. "Traditional community leaders had access to knowledge transmitted from their ancestors about nature, including medicinal plants, animal behavior, all aspects of the environment supporting their lives. That knowledge might not have been publicly shared or widely available, only accessible to select members of the community. But through traditional practices it was connected to decision-making as part of the larger fabric of how that community lived and related to nature. I like to think of traditional leaders as the 'community GIS' for the people, because they would integrate layers of knowledge of plants, animals, and other resources, and, through spirits and rituals, unlock that knowledge to guide community decisions and change people's behavior."

By contrast, Lilian says, "Consider what's happening with modern ways of learning and decision-making. To become a doctor, you go to medical school. To become a botanist, you study plants at the university. To become a hydrologist, you might study geoscience or engineering or hydrology. We're so specialized. As a result, we have unprecedented, in-depth understanding and knowledge of the individual disciplines. We have different government agencies managing individual sectors. But the traditional leaders and the respect for plants, animals, and the shared environment is not there anymore to bring all this knowledge together and integrate it to inform wiser decisions. These key knowledge-holders, along with the rituals, traditions, and practices helping us connect with each other and with larger ecosystems, are becoming a thing of the past."

As Joseline Nyangoma noted in an earlier chapter, our current data and knowledge are fragmented, with different software systems, computers, and servers owned by different institutions and NGOs. A lack of data standards, local resources, and capacities is also keeping this information compartmentalized. "One promise of GIS is bringing it all together, if we do it right, with the people of the place," Lilian says. "GIS enables us to overlay, connect, and integrate individual layers of information to get new insights through the power of geography. Different layers of information that have become segmented can be unlocked and organized into one pool of knowledge to achieve wiser decisions. In this way, GIS could, in some respects, take on the role of a traditional holistic knowledge-keeper, and beyond that, serve as the intelligent nervous system of the planet, as Jack Dangermond has suggested. Not only to help us connect, understand, and make sense of the world but also to repair our relationship with it."

Because the relationships required for activities like participatory mapping are founded on trust and inclusion, JGI and other community-conservation practitioners must be sensitive to the way traditional and contemporary beliefs and the physical and spiritual worlds are intertwined in the everyday lives of these communities. Discounting any aspect of their collective reality can lead to miscommunication and loss of trust. Information-sharing, including traditional ecological knowledge, is essential if conservation practitioners are to engage and empower local communities to own and drive conservation decisions on their lands.

To illustrate the value of traditional ecological knowledge, Lilian tells a story about a village east of Gombe. Because the village is densely populated, most of the habitat had been lost, so Lilian had low expectations of what was possible and what level of forest restoration or conservation could be achieved on the village land. Nevertheless, Lilian, Kashula, and other JGI colleagues conducted a community mapping exercise using printouts of the satellite imagery of the village land.

"The minute we put these satellite image maps on the wall, the private forest owners started locating their forests," Lilian recalls. "Then we started comparing images from different years and realized that, despite high population density, the village had indeed been successful in restoring and protecting some of their forests. The private forest owners had decided, independently, to restore and protect their own forest patches. These patches are small and not necessarily connected to each other but could have a positive impact on local biodiversity."

Later, Lilian says, they stopped by the house of a village forest

Mapping local knowledge through participatory interpretation of Maxar satellite images with village elders and observers. *The Jane Goodall Institute, Lilian Pintea.*

monitor to pick him up to visit the forest reserves. "A few elders were resting outside the house with children playing nearby. In front of them was a basket of chopped tree bark drying in the sun. I recognized that it was a traditional medicine, so I greeted them and respectfully asked if they could show on the satellite image map where this traditional medicine was coming from." The elders agreed and immediately began locating features of their village on the map, including a private forest patch where they sourced the traditional medicine.

"Then the forest monitor pointed to the forest he had been patrolling regularly—the same forest where they had been collecting the medicinal tree bark. We traveled to that private forest, and, after mapping a small stream running through it, I asked if the monitor had recorded

Kakombe waterfall in Gombe National Park. The site has multiple meanings to residents and is considered sacred. Local communities are allowed to visit the site on special occasions as part of their traditional beliefs. *The Jane Goodall Institute, Lilian Pintea.*

any wildlife in the reserve and where the most recent sightings were. He said, 'I can show you my guardians, who are helping to protect my forests. They are snakes—puff adders.' Not really knowing if I had my translation wrong, I wasn't convinced. He said, 'No, really, I can show you.' We were still thinking he was probably talking about another harmless reptile, maybe a monitor lizard, so I pulled up a picture of a puff adder on my phone to show him. He said, 'Yes, yes, that's it.' Immediately everyone started running away," says Lilian with a laugh, reliving the moment of showing an image of a puff adder (a venomous viper species).

"I said, 'Wait a minute, it seems you have a special relationship with the snakes—could you share more about it?' The forest monitor said, 'Well, the elder you met who had the bark medicine drying back at the house, he is my great-uncle. He is the owner of the forest, and our great-great-uncle passed it to him. This is a family forest, and now it's my turn to learn about the forest and protect it, and I'm doing that by using the JGI mobile phone to monitor, record, and report illegal activities. But I also learned from my great-uncle, so I invite the help of these five puff adders, which are at the border of the reserve. I give them offerings, a little bit of millet and other grain, and I sprinkle it in front of their burrows. In return, they are protecting and guarding my forest, our forest."

Although puff adders do not eat grain offerings, small mammals such as rodents do, and luring a puff adder's prey to the burrow entrance may help to keep these snakes in the area. Again, traditional ecological knowledge and cultural heritage have a significant role to play in conservation, as the older ways of living for virtually all indigenous people are woven into their environment in ways that manifest as mutualistic relationships—relationships that entail a sense of belonging to the landscape, rather than owning it. For Lilian, what's significant is that the community forest monitor, who was selected by the village government, is not only acquiring and using new tools (for example, JGI smartphones and ArcGIS Survey123) but also integrating them with his ancestral knowledge.

"We're still figuring this out," says Lilian, referring to the challenges of scaling up such hyper-localized, contextual knowledge, "but using both puff adders and smartphones—why not?" This is what it means to incorporate a completely different way of knowing and combine it with science and geospatial technology to create usable, science-based, and community-owned programs. Whether it's a question of an outsider navigating unfamiliar belief systems or a young man trying to keep hold of his family heritage by combining ancient and modern knowledge, the Tacare

Participatory mapping with local communities in Bitale village, using Maxar satellite images and the Tacare approach. *The Jane Goodall Institute. Lilian Pintea.*

approach is about encouraging local people to own their future through community participation. All types of knowledge have their place, Lilian believes, because "there are many stakeholders in this world, and we're all beneficiaries of a healthy environment."

Conclusion

W hen it began in Tanzania in 1994, the TACARE project was inspired by Jane Goodall's message, "Every individual matters," and her question, "How can we save the chimps if the people outside the forest are struggling to survive?" In the decades since, the project has evolved into a proven nature-based development approach aimed at empowering local communities, leaders, and individuals to own the process of decision-making and the management of their natural resources. What is now known as the Tacare approach has been introduced to Uganda, Republic of the Congo, Senegal, and the DRC, with future goals to extend its reach globally beyond JGI and the ranges of the great apes in Africa.

Jane Goodall was one of the first conservationists to recognize the need to put local people at the center of conservation. Working hand in hand with communities living in or near habitat of the great apes, JGI's community-led approach has become even more holistic over time, using different ways of knowing, new participatory methods, conservation planning tools, and geospatial technologies. Although the concept of community-based conservation is widely accepted today, the Tacare approach is distinct. Tacare not only engages and works with local people and institutions but strives to assure that they own and drive development and conservation decisions in their landscapes as they address their own needs, the needs of wildlife, and the needs of their shared environment.

The Tacare approach recognizes that local people are the most connected to and dependent on healthy landscapes and ecosystem services, having survived and adapted over time through indigenous and traditional knowledge. Tacare also acknowledges that local people are the most impacted and vulnerable when ecosystem services disappear. For these reasons, JGI believes that local communities are the best stewards of their own environments and that every community member can make a difference every day.

Tacare's journey from a reforestation project to a broader approach has not been a linear one. Indeed, more than 25 years of trial and error have refined Tacare to where it is today, with still many challenges ahead. Through the journey of discovery described in this book, the fundamentals of the Tacare approach are now established, yet their application continues to adapt and evolve, responding to new environmental and social problems.

Tacare rests on the fundamental phases of Engage, Listen, Understand, and Facilitate—with a fifth, Step Back, in cases where the community elects not to participate or where true sustainable practices have been put in place. These phases are not always sequential, nor are they linear. Rather, they form a web of intersecting interactions and feedback loops, all of which are subject to modification at the hands of those for whom Tacare is designed—local people and institutions.

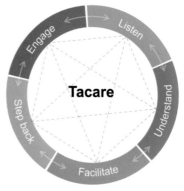

The Tacare conceptual model showing a web of interactions and feedback loops connecting the five stages of the process. *The Jane Goodall Institute.*

Engage

Tacare starts with proper engagement. The Engage phase could be as simple as an invitation by the community, or a meeting with community leaders and local governments, to begin a dialogue. If the community is not ready, the Tacare way is to step back, respecting and understanding that it might take years before a community is ready and interested to move to the next stage. Tacare does not dictate the terms of conservation to the local people; it shares and offers access to knowledge, technologies, and tools available for nature-based development and conservation. The communities themselves then choose which path they wish to follow.

Listen

The Listen phase is one of Tacare's key guiding principles, a way to engage people's hearts and minds with compassion throughout the process. Active listening means listening not only to the needs and concerns of local people but also to their insights, values, beliefs, and ways of knowing. Today, the listening phase is supported by conceptual mapping and the participatory interpretation of high-resolution satellite imagery to help explore and document people's perceptions, knowledge of their environment, and development needs. Over the years, Tacare staff have learned that listening is the key to community buy-in—building mutual partnerships and trust—and essential for the success of any development and conservation activities.

Understand

The Understand phase of the Tacare approach is developing a common and shared understanding of the problem and potential solutions. Conservation is inherently spatial and complex. To understand, Tacare uses geospatial technologies, following geodesign principles with the people of the place to combine traditional and scientific knowledge and develop conceptual models of how the needs of animals, people, and the environment are interconnected. In this way, the process works to define the major problems and what can be done about them. In the early days of Tacare, JGI learned that people cared about access to clean water, food, health, education, and job opportunities but did not really think about or speak for the needs of chimpanzees, other wildlife, forests, and ecosystems. Clearly, it was necessary to help people understand how the health of the ecosystem—which depends on biodiversity—is connected to their own well-being, their livelihoods, and their future.

JGI understood the need to complement the initial Listen phase with a conservation action planning approach that could help answer these questions:

- How are species and habitats doing in this community and larger ecosystem, and what are the major threats to their survival?

- What actions are needed to address those priority threats?

- Are conservation actions effective in addressing the most important threats?

To answer such questions, JGI uses Open Standards for the Practice of Conservation (Conservation Standards, for short) as part of the Tacare approach. Conservation Standards is a science-based and collaborative planning approach developed by the Conservation Measures Partnership that uses adaptive management to help focus conservation decisions and actions on clearly defined objectives and measures success in a manner that enables adaptation and learning over time.

Facilitate

The next phase in the Tacare approach—Facilitate—empowers individuals, community leaders, members, and groups to implement their local solutions themselves. This phase includes activities such as building local capacity and securing resources to support community plans. Local solutions are holistic and range from land use planning and forest monitoring using mobile technologies to improved farming and agroforestry practices, water management and marketing skills, improved primary health care and children's education, microcredit opportunities for women, and

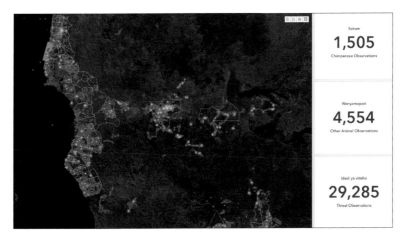

Example of a community village forest monitoring dashboard developed as part of the Western Tanzania Decision Support and Alert System, using ArcGIS Survey123 and ArcGIS Experience Builder. *The Jane Goodall Institute, Lilian Pintea.*

scholarships for girls. Such initiatives all include measures to help communities live in harmony with their environment.

As these needs are addressed and solutions implemented, conservation of the surrounding ecosystems and other species begins to improve. Ecosystem recovery becomes the byproduct of these nature-based initiatives, designed to help the people first. After all, the loss of a species or its habitat as the result of human activities is not only a problem in itself but an indicator of a larger breakdown of healthy ecosystems and the relationship between people and nature. It's also often the manifestation of a growing human population, poverty, and loss of natural resources, forcing people to survive by any means necessary. For this reason, imposing initiatives aimed solely at a single outcome, like chimp conservation, is unlikely to work in the long term. Tacare thus combines short- and long-term strategies to address the cause, not only the symptoms. The approach supports a multitude of initiatives at the village level, all of which vary according to suitability, relevance, and above all, community adoption.

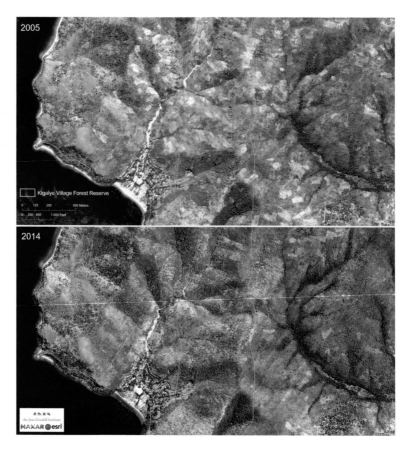

Natural regeneration in Kigalye village between 2005 (*top image*) and 2014, showing improvements in tree cover on steep slopes, streams, and watersheds, derived from Maxar satellite images. *The Jane Goodall Institute, Lilian Pintea.*

Step back

Because Tacare is offered, not imposed, the local people choose to participate and then freely own and drive the programs. However, if a particular village does not wish to be engaged by JGI or adopt its developmental approaches, JGI withdraws from its engagement. Yet this withdrawal always comes with an open invitation to the local people should they later

Roots & Shoots students and teachers map their school environments using ArcGIS Tacare community mapping apps at the Jane Goodall Nature Center in Pugu, Tanzania, with support from Esri Eastern Africa. *The Jane Goodall Institute, Japhet Jonas Mwanang'ombe.*

choose to re-engage and become custodians of these initiatives. In many cases, re-engagement is the eventual outcome.

Ultimately, the Tacare approach is about empathy and a call for action to bring more resources and attention to people living at the "last mile" of the conservation cycle. When local people are heard, understood, and involved in every step, they take ownership of generating creative solutions to ongoing challenges, such as resource use, health, and education, while developing a deeper understanding of their own place within the ecology of a region. Tacare also aims to foster the precious traditional ecological knowledge held by community members, recognizing its value in informing land and resource use.

What distinguishes the Tacare approach is that the local people are offered the option to participate in the programs and then own and drive the programs themselves. Tacare answers first to the local people, ahead of

Example of a community map and dashboard facilitated by Roots & Shoots Tanzania, using ArcGIS apps customized by Esri Eastern Africa. *The Jane Goodall Institute, Lilian Pintea.*

donors, international advisers, and academics, even ahead of the organization's owns needs to meet program targets. Tacare's message is that such decolonization of conservation is urgently needed as the most effective way to ensure engagement, independence, and ownership, and hence to achieve long-term conservation. Change takes time, but if seeds are sown under the right conditions, eventually they will sprout.

Although it is a distinct branch of JGI, the Roots & Shoots program, which began in 1991, is an integral component of today's Tacare approach. Aimed at reaching global youth and empowering them to act on topics of conservation and environmental sustainability, Roots & Shoots shares Tacare's premise of "change from within" and provides another platform from which to launch Tacare among community members. It provides students with information about different plant and animal species and their importance in the ecosystem, helping young people understand that animals are sentient beings, not just resources to be exploited.

Local Voices, Local Choices presents a rich collection of stories from the people on the ground, exploring the evolution of the Tacare approach through their own experiences. Structured chronologically, from JGI's origins in 1960, these chapters weave together the voices of the very people that make Tacare possible—not only those of the dedicated JGI staff and program partners but also, and equally, those of local people. In telling their stories, these participants have transparently described the program's ups and downs and successes and failures, so that the Tacare approach may be understood and, as Jane Goodall hopes, adapted and scaled anywhere in the world where there is a place for genuine community-led conservation.

Contributors

Dr. Lilian Pintea

Vice President, Conservation Science, the Jane Goodall Institute, USA

Dr. Lilian Pintea. *The Jane Goodall Institute.*

Lilian is recognized as a pioneer in applying innovative geospatial technologies to conservation. He brings more than 25 years of experience in applying satellite imagery, GIS, cloud, and mobile technologies to the job of conserving chimpanzees, other wildlife, and their vanishing habitats. As vice president of conservation science at JGI, Dr. Pintea and his team oversee science activities and functions at the institute, supporting departments and country offices and bringing targeted research, analysis, and technological innovation to support JGI's mission. Lilian has contributed to the development of the Tacare approach since 2000. He secured funding, managed and contributed to the concept and design of *Local Voices, Local*

Choices: The Tacare Approach to Community-Led Conservation, provided maps and figures for this book, and wrote and coedited the final draft.

Adam Bean. *The Jane Goodall Institute.*

Adam Bean

Manager, Conservation Science, the Jane Goodall Institute, USA

Adam has worked for JGI since 2019. Before this, he acquired 20 years of experience from a variety of wildlife conservation positions in zoology—specifically the fields of captive wildlife management, herpetology, and animal behavior—and worked in field-based conservation roles on various threatened species programs in Australia, Namibia, and Nigeria. Adam wrote the narrative text for this book and conducted the book's content research throughout East Africa, interviewing both current and former JGI staff, government officials, and local community members to produce the first publication draft. Adam also curated the book's photo content and coedited the final draft.

About Esri Press

A t Esri Press, our mission is to inform, inspire, and teach professionals, students, educators, and the public about GIS by developing print and digital publications. Our goal is to increase the adoption of ArcGIS and to support the vision and brand of Esri. We strive to be the leader in publishing great GIS books, and we are dedicated to improving the work and lives of our global community of users, authors, and colleagues.

Acquisitions

Stacy Krieg
Claudia Naber
Alycia Tornetta
Craig Carpenter
Jenefer Shute

Editorial

Carolyn Schatz
Mark Henry
David Oberman

Production

Monica McGregor
Victoria Roberts

Marketing

Mike Livingston
Sasha Gallardo
Beth Bauler

Contributors

Christian Harder
Matt Artz
Keith Mann

Business

Catherine Ortiz
Jon Carter
Jason Childs

For information on Esri Press books and resources, visit our website at esri.com/en-us/esri-press.